PHYSICS AT SEVENTEENTH
AND EIGHTEENTH-CENTURY LEIDEN:
PHILOSOPHY AND THE NEW SCIENCE
IN THE UNIVERSITY

ARCHIVES INTERNATIONALES D'HISTOIRE DES IDEES

INTERNATIONAL ARCHIVES OF THE HISTORY OF IDEAS

Series Minor

11

EDWARD G. RUESTOW

PHYSICS AT SEVENTEENTH
AND EIGHTEENTH-CENTURY LEIDEN:
PHILOSOPHY AND THE NEW SCIENCE
IN THE UNIVERSITY

Directors: P. Dibon (Paris) and R. Popkin (Univ. of California, La Jolla)
Editorial Board: J. Aubin (Paris); J. Collins (St. Louis Univ.); P. Costabel (Paris); A. Crombie (Oxford); I. Dambska (Cracow); H. de la Fontaine-Verwey (Amsterdam); H. Gadamer (Heidelberg); H. Gouhier (Paris); T. Gregory (Rome); T. E. Jessop (Hull); P. O. Kristeller (Columbia Univ.); Elisabeth Labrousse (Paris); A. Lossky (Los Angeles); S. Lindroth (Upsala); P. Mesnard (Tours); J. Orcibal (Paris); I. S. Révah † (Paris); J. Roger (Paris); H. Rowen (Rutgers Univ., N.Y.); G. Sebba (Emory Univ., Atlanta); R. Shackleton (Oxford); J. Tans (Groningen); G. Tonelli (Binghamton, N.Y.).

PHYSICS AT SEVENTEENTH AND EIGHTEENTH-CENTURY LEIDEN: PHILOSOPHY AND THE NEW SCIENCE IN THE UNIVERSITY

by

EDWARD G. RUESTOW

MARTINUS NIJHOFF / THE HAGUE / 1973

Dedicated to

CLARA FRANCES AND PAUL ERNEST RUESTOW

with much affection

© 1973 by Martinus Nijhoff, The Hague, Netherlands
All rights reserved, including the right to translate or to
reproduce this book or parts thereof in any form

ISBN 90 247 1557 1

PRINTED IN THE NETHERLANDS

TABLE OF CONTENTS

Acknowledgements	VII
CHAPTER I. INTRODUCTION: A NEW UNIVERSITY AND THE CHALLENGE OF THE NEW SCIENCE	1
CHAPTER II. FRANCO BURGERSDIJCK: LATE SCHOLASTICISM AT LEIDEN	14
CHAPTER III. TUMULT OVER CARTESIANISM	34
CHAPTER IV. JOANNES DE RAEY: THE INTRODUCTION OF CARTESIAN PHYSICS AT LEIDEN	61
CHAPTER V. PASSING CRISES, ENDURING DISAGREEMENT	73
CHAPTER VI. THE PRACTICE OF PHILOSOPHY	89
CHAPTER VII. 'S GRAVESANDE AND MUSSCHENBROEK: NEWTONIANISM AT LEIDEN	113
CHAPTER VIII. CONCLUSION: SCIENCE, PHILOSOPHY AND PEDAGOGY	140
Selected Bibliography	155

ACKNOWLEDGEMENTS

Begun as a dissertation at Indiana University, this study has remained heavily indebted to the sympathetic criticism of Richard S. Westfall. The suggestions that he has offered that have contributed to the shaping of this book could never be completely enumerated, and I can only acknowledge that I have taken great advantage of his continuing interest. I am also beholden to Paul Miller of the University of Colorado, whose critical reading of Chapter II did much to rectify my understanding of the Aristotelian and scholastic traditions. If that understanding may in places still seem wanting, it can be attributed only to my own obstinacy. I owe particular gratitude as well to Prof. Dr. S. van der Woude, Librarian of the University Library of Amsterdam, for permitting me to peruse the catalogue he is compiling of Dutch university disputations in European libraries, and to P. W. Tiele of the University of Leiden Library staff, who was always a source of the most ready and willing assistance. Finally, the financial support that allowed me to carry out this study was provided largely by a Graduate Fellowship awarded by Indiana University in 1966-7, an NDEA-Related Fulbright-Hays Grant the following year, and an Initiation Faculty Fellowship from the University of Colorado for the summer of 1971.

CHAPTER I

INTRODUCTION: A NEW UNIVERSITY AND
THE CHALLENGE OF THE NEW SCIENCE

Despite the recent and continuing controversy concerning the proper role of the universities, it is still widely agreed that among their functions, if not their principal function, is the transmission of a cultural heritage. It is also now a commonplace, however, that the universities have often been slow to recognize when the abridged heritage they cared for needed to be altered to make room for new intellectual pursuits cultivated outside their walls. Long an outstanding case in point has been the tardiness of the universities in embracing the new natural science developed during the course of the seventeenth century. Frequently, indeed, the universities proved bastions of conservative resistance, and, generally, they remained apathetic at best for another two centuries, compelling the devotees of the new science to create their own new institutions of higher learning.[1]

The response of the universities initially reflected, in large part, the anxiety of the organized churches, which looked upon the emerging science as a threat to sound doctrine and upon the universities as schools for the clergy. Behind the churches also loomed the state governments, growing ominously in power and often increasingly opposed, in their turn, to any distraction from the services they, too, now expected from the universities. In addition to the external pressures brought to bear on the schools by the more powerful institutions of society, however, there were as well the troubling philosophical difficulties that the new science itself created for educated Europeans. Such difficulties were especially acute within the universities, where they were further augmented by patterns of thought deeply rooted in the academic tradition and incompatible with the procedures and attitudes of the new science. Indeed, it may be asked to what extent the conservatism of the universities was the

---

[1] See Martha Ornstein, *The Rôle of the Scientific Societies in the Seventeenth Century* (3rd. ed.; Chicago: The University of Chicago Press, 1938).

result of the attitudes characteristic of the small group of permanent residents at the schools, the academic scholars.

This conservatism, however, was not everywhere equally efficacious. In the sixteenth century, the universities of northern Italy, Padua above all, had nurtured an intellectual ferment of considerable significance to the rise of the new science, and they continued to be penetrated by the influence of that science throughout the seventeenth century. The University of Oxford momentarily played host to leading members of the English scientific community during the Commonwealth period, and Cambridge was shortly to boast the genius of Isaac Newton. Indeed, a small number of the one-hundred-odd universities in Europe strove more or less purposefully to come to grips with the new science and to incorporate facets of it, at least, within the body of learning for which they held themselves responsible.[2] Among the most notable of these more progressive schools must be included the University of Leiden, recently founded by the Lowlanders in revolt against the King of Spain, Philip II.

The doors of the University of Leiden had first opened, to be sure, in the midst of rebellion, and had been forced open, as it were, by rumors of peace. In 1572, the revolt, with the Calvinists now clearly in the van, acquired what was to prove an enduring foothold in the maritime provinces of Holland and Zeeland. To strengthen the solidarity of the rebels and to provide for a trained and educated Calvinist clergy, William of Orange, the aristocratic symbol of the revolt, proposed the founding of a university for the rebellious provinces, for the University of Louvain, long the pride of the Netherlands, remained in the hands of the Spanish and a stronghold of the Counter-Reformation. Reports in 1574 that peace negotiations were being planned forced the hasty realization of William's project, since Philip II would hardly have countenanced the establishment of a Protestant university once his authority had been restored.[3] In tribute to her heroic resistance in the face of a Spanish siege, the city of Leiden was decided upon as the seat of the new academy. Though its building and initial revenues were also to be provided, as was the case with so many Reformation universities, from properties once belonging

---

[2] For a list of universities founded in medieval and early modern Europe, see the appendix of Arturo Castiglioni's *A History of Medicine* (trans. E. B. Krumbhaar; New York: A. Knopf, 1941), pp. 962-3.

[3] In 1562, Philip himself had already founded a new university in the Lowlands, the University of Douai, with the hope, precisely, of striking a blow against the influence of the Calvinist Academy at Geneva. Léon Van der Essen, *Une institution d'enseignement supérieur sous l'ancien régime: L'Université de Louvain* (Bruxelles et Paris: Vromant et Co., 1921), p. 30.

to the Catholic Church, the academy was officially founded by William in 1575 in Philip's name.[4]

The anticipated peace, however, did not materialize, and seven years after he had ostensibly established the new university, Philip II issued a prohibition against it, all his subjects who studied there being henceforth suspected of heresy and considered unfit for any offices or dignities in his realms.[5] Interrupted by twelve years of truce between 1609 and 1621, the war between the rebellious provinces and the Spanish crown continued until the Peace of Münster in 1648. Only then would Spain accept what had long been a reality, the existence of an independent republic, the United Provinces, encompassing the northern seven of the original seventeen provinces of the Lowlands.

The university founded at Leiden nearly three quarters of a century earlier had by now been joined by four other new universities within the republic – Franeker, founded in 1585; Groningen, 1614; Utrecht, 1636; and Harderwijk, 1648 – and at least six other institutions with higher faculties but no power of granting degrees. Leiden never relinquished her preeminence, however, and continued in the late seventeenth and early eighteenth century to attract more students than her four sister universities combined.[6] During this period, she won for herself undisputed recognition as one of the leading universities of Europe.

Hurried into existence, Leiden had first opened her doors to a total student enrollment of two, but by the beginning of the seventeenth century, new registrants were numbering approximately 150 a year and were increasing steadily.[7] Half a century later, the student body numbered about one thousand, still significantly below the two or three thousand at Oxford and Cambridge – much less the massive twenty thousand that were to be found, it had been claimed, at sixteenth-century Paris – but nonetheless large at a time when three or four hundred made for a fair-sized university.[8] By the end of the century, when new registrants were

---

[4] See Pieter Geyl, *The Revolt of the Netherlands, 1555-1609* (London: Ernest Benn Limited, 1958), pp. 136-8.

[5] *Bronnen tot de Geschiedenis der Leidsche Universiteit*, ed. P. C. Molhuysen ('s-Gravenhage: Martinus Nijhoff, 1913-24), Vol. I, p. 101*.

[6] A. C. J. de Vrankrijker, *Vier Eeuwen Nederlandsch Studentenleven* (Voorburg: Uitgeverij Boot N.V., [1939]), p. 44.

[7] Matthijs Siegenbeek, *Geschiedenis der Leidsche Hoogeschool, van hare Oprigting in den Jare 1575, tot het Jaar 1825* (Leiden: S. en J. Luchtmans, 1829-32), Vol. I, p. 62.

[8] *Ibid.*, p. 153. Charles Edward Mallet, *History of the University of Oxford* (London: Methuen and Co., Ltd., 1924-7), Vol. II, p. 243. James Bass Mullinger, *The University of Cambridge* (Cambridge: The University Press, 1873-1911), Vol. II, pp. 214 and 574. Samuel Eliot Morison, *The Founding of Harvard College* (Cambridge, Mass.: Harvard University Press, 1935), p. 42. Aristide Douarche, *L'Université de Paris et les Jésuites* (Genève: Slatkine Reprints, 1970), p. 311.

averaging over three hundred a year, the student body was reaching its peak, probably climbing above twelve hundred.[9] There was a slight falling off in the first years of the eighteenth century, an even slighter rally, and then, after the first quarter of the century, a gradual but steady decline in numbers that reflected the rise of the German and Scottish universities to the east and west.

The foreign student population, equal at times to the number of Dutch students, constituted a substantial but highly mobile segment of Leiden's student body throughout the period.[10] The great majority came from northern and eastern Europe, primarily Germany, England and Scotland, and their sojourn at Leiden, frequently for pleasure alone or a degree to cap studies completed elsewhere, was often brief. Essential to Leiden's appeal among them was the religious toleration that characterized her admission policy in a time marked by confessional exclusiveness throughout Europe. Conceived initially as a specifically Calvinist academy, Leiden's statutes had originally demanded an oath from the students to adhere, while students at Leiden, to no other doctrine than that professed there.[11] Within three years of its founding, however, following student agitation, the requirement for such an oath was revoked for all students except those studying theology, and the States of Holland and Westfriesland invited everyone, "of whatever state, condition, religion or quality he may be," to come study at Leiden in freedom and security.[12] What remained was an oath of obedience to the statutes and university authorities, an oath itself much abused in that age of student turbulence.[13] The oath taken by the faculty was limited to a similar commitment of obedience, but at this level within the university, the

---

George Clark, *The Seventeenth Century* (2nd ed.; New York: Oxford University Press, 1961), p. 288.

[9] de Vrankrijker, *Vier Eeuwen*, p. 44. Siegenbeek, *Geschiedenis der Leidsche Hoogeschool*, Vol. I, pp. 200, 287 and 288. De Vrankrijker stresses that enrollment figures cannot be used to determine the precise size of the student body, for the figures give no idea of how long students remained at Leiden and often include non-students seeking to profit from the university privileges (p. 42). A Silesian student entering the university in 1665 asserted that the student population at the time was as high as three thousand (G. D. J. Schotel, *De Academie te Leiden in*

[10] Cf. Schotel, *op. cit.*, pp. 264-84, and de Vrankrijker, *op. cit.*, p. 46 ff.

[11] *Bronnen*, Vol. I, p. 32*.

[12] *Ibid.*, pp. 56*-57*.

[13] *Ibid.*, p. 315*.

[14] *Ibid.*, 183*. The faculty at Leiden would prove, with time, inaccessible to Catholic scholars (L. J. Rogier, *Geschiedenis van het Katholicisme in Noord-Nederland in de 16e en de 17e Eeuw* [Amsterdam, 1946], Vol. II, p. 718), and Burchardus de Volder, a Mennonite, would find it necessary to convert to the Reformed Church before assuming his post in the faculty of philosophy in the late seventeenth century (see below, p. 74).

pressure towards religious conformity would in the long run prevail.[14] The student body, nonetheless, continued to be characterized by a wide diversity of faiths, including Catholicism as well as Eastern Orthodoxy, Judaism and Socinianism.[15]

The organization of the faculty conformed to the time-honored framework of the universities of northern Europe, with three higher faculties of theology, medicine and law and a lower preliminary faculty of arts or philosophy. This latter faculty was responsible for the major philosophical disciplines – logic, physics, ethics and, though not officially taught at Leiden until the mid-seventeenth century, metaphysics – as well as Latin, Greek, mathematics, Hebrew, history and, at Leiden, Arabic. It was as a part of the offerings of this faculty that classical and philological studies were to acquire particular importance at Leiden during the early decades of her existence.

By the middle of the seventeenth century, Leiden's faculty had grown from eleven to twenty members, and, supplemented by temporary "lectors" and private teachers, it remained approximately that size into the eighteenth century.[16] The individual faculty members were selected and appointed by a board of curators (or, more precisely, of curators and burgemeesters) that reflected the dual provincial and municipal character of the academy. The board was composed of seven members and a secretary, four of the members being the burgemeesters of the city, who changed yearly, and the other three, life-time appointees of the States of Holland and Westfriesland.[17] One of the appointees of the States was a nobleman, while the other two were usually drawn from among the governing officials of the other major cities of Holland.[18] The senior professors, the professors "ordinary," together constituted the academic senate and, as such, annually elected from among themselves the rector, who wielded great influence in administrative matters. But the board of curators remained the real governing body of the university. It not only appointed the faculty and determined their salaries, but, together with the rector and other academic officers, had the power to prescribe the books and material to be taught.[19]

Dedicated to winning a reputation for their new university, the curators were marked from the beginning by their competitiveness, seeking to embellish the young academy with the names of renowned scholars.

[15] Rogier, *Ibid.*, pp. 32 and 60. Nicholas Hans, "Holland in the Eighteenth Century – *Verlichting*," Paedagogica Historica, Vol. V (1965), p. 23.
[16] Schotel, *De Academie te Leiden*, pp. 242-3.
[17] *Bronnen*, Vol. I, p. xii.
[18] Schotel, *De Academie te Leiden*, p. 210.
[19] *Bronnen*, Vol. I, p. 29*.

Three years after Leiden was founded, the celebrated classicist Justus Lipsius was enticed away from Louvain, and when he left thirteen years later to return to Louvain and Catholicism, the King of France himself was induced to assist in persuading the no less famous classicist and historian, Josephus Justus Scaliger – the "bottomless pit of erudition" [20] – to join the aspiring school on virtually his own terms. In the same year, 1593, the lame and aged Carolus Clusius, one of Europe's best known botanists, was likewise lured to Leiden with a handsome stipend and little to do beyond allowing his name to enhance the reputation of the university and its newly-established botanical garden.

Throughout the years, the curators were generally to prove sympathetic to the desires of the faculty for the expansion of facilities and innovations in the university's offerings. In 1587, Leiden had joined the handful of sixteenth-century universities, most of them in Italy, which had begun expanding their medical instruction by the establishment of botanical gardens.[21] Constantly enriched by the world-roving Dutch sea captains, the garden at Leiden would contain by the end of the seventeenth century what was probably the most richly varied collection of living plants in the world.[22] An anatomical theatre was also constructed in 1597, where Petrus Pauw, who had studied under Fabricius at Padua, continued the efforts of the great Italian anatomists. In 1633, the first observatory – strictly for purposes of teaching, however – ever attached to a public institution in Europe was erected on the roof of the academy building, and in 1669, a chemical laboratory, likewise perhaps the first such university laboratory, was also established.[23] Six years later, one of Europe's earliest demonstration halls for experimental physics was equipped at Leiden and a pioneering course in that subject begun.[24] Leiden's most consequential initiative, however, both for her own fame and the future development of advanced academic training, was the inauguration of clinical instruction in medicine in 1638, an innovation with

[20] Quoted by Morison, *The Founding of Harvard College*, p. 141.
[21] Just Emile Kroon, *Bijdragen tot de Geschiedenis van het geneeskundig Onderwijs aan de Leidsche Universiteit 1575-1625* (Leiden: S. C. van Doesburgh, 1911), p. 69.
[22] William Thomas Stearn, "The Influence of Leyden on Botany in the Seventeenth and Eighteenth Centuries," *Early Leyden Botany* (Universitaire Pers Leiden; Assen: Van Gorcum en Comp. N.V., 1961), p. 21.
[23] Willem de Sitter, *Short History of the Observatory of the University at Leiden, 1633-1933* (Haarlem: Joh. Enschede en zonen, [1933]), pp. 8-10. *Bronnen*, Vol. III, pp. 227-30 and 243. Martha Ornstein, though aware of the earlier chemical laboratory at Leiden, nonetheless describes that built at the University of Altdorf in 1682 as "the first chemical university laboratory of the world" (*The Rôle of the Scientific Societies*, pp. 231 and 252).
[24] See below, p. 96 ff.

which Padua had experimented in the previous century, but with little or no lasting effect.[25] Firmly established as a practice at Leiden, clinical instruction became a permanent and essential part of medical education.

As was characteristic of many universities in that age of intensive state-building, it was the faculty of law that claimed the largest number of students, but the period of Leiden's greatest international influence owed more to the renown of her medical faculty.[26] That faculty had not been without its distinguished members earlier in the seventeenth century, but it acquired particular eminence after Franciscus de le Boe Sylvius joined its ranks in 1658. Having himself been briefly a student at Leiden, he had first returned in 1638. Remaining for perhaps three years, he was granted the right to offer private lessons and, demonstrating with dogs, first won the members of the regular faculty to Harvey's theory of the circulation of the blood. Notable among the converts was Johan de Wale, who, initiating his own series of experiments, became one of the foremost proponents of the theory.[27] Leiden, indeed, may contest with the University of Jena the distinction of having been the first continental university to have taught Harvey's revolutionary doctrine.[28]

When Sylvius again returned two decades later as a regular member of the faculty, he intensified the clinical instruction, stimulated physiological investigation, and began Leiden's first instruction in chemistry.[29] He became, as well, the founder and chief exponent of the century's second great school of speculative physiology, the iatrochemists, who stressed chemical operations, as opposed to the pure mechanism of the iatromechanists, as the basis of the physiological processes of the body. Characterized by ill-founded theorizing about "fermentation," however, his doctrines of physiological chemistry were hotly contested at Leiden itself.

The prestige of the medical faculty at Leiden reached its apogee during the professorate of Hermannus Boerhaave, in his day the most celebrated name in medicine.[30] Appointed lector in 1701 and professor

[25] Theodor Puschmann, *A History of Medical Education*, ed. and trans. Evan H. Hare (London: H. K. Lewis, 1891), pp. 410-1.

[26] Schotel, *De Academie te Leiden*, pp. 146-7.

[27] See A. H. Israëls, "De Verdiensten der Nederlanders in het Verspreiden en Uitbreiden der Harveyaansche Ontdekking," *Nederlandsch Tijdschrift voor Geneeskunde* (tevens orgaan der Nederlandsche Maatschappij tot Bevordering der Geneeskunst), Vol. IV (1860), pp. 361-73.

[28] *Ibid.*, p. 364.

[29] W. P. Jorissen, *Het chemisch (thans anorganisch chemisch) Laboratorium der Universiteit te Leiden van 1859-1909 en de chemische Laboratoria dier Universiteit vóór dat Tijdvak en hen, die er in doceerden* (Leiden: A. W. Sijthoff's Uitgevers-Maatschappij, 1909), p. 9.

[30] Stephen d'Irsay, *Histoire des universités françaises et étrangères des origines à nos jours* (Paris: Éditions Auguste Picard, 1933-5), Vol. II, p. 81.

in 1709, he continued teaching at Leiden until the year of his death, 1738, having taught not only theoretical medicine but clinical medicine, botany, and chemistry as well.[31] His fame rested on his greatness as a teacher and systematizer of the chaotic body of fact and opinion embraced within eighteenth-century medical thought, and through his students his influence was to dominate the progressive development of medical instruction throughout eighteenth-century Europe.[32]

By the time of Boerhaave's professorate, indeed, the curators of the University of Leiden might well have claimed that theirs was now the outstanding university of Europe, for the rise of Leiden had taken place against a background of general stagnation and decline elsewhere. The great Italian universities, it is true, had remained vigorous centers of activity, but they had gradually lost the preeminence they had enjoyed in European intellectual life in the sixteenth century, and in some respects, particularly in the field of medicine, Leiden could be regarded as their successor.[33] The sixteenth century had also been an era of brilliance for the universities of Spain, but the following century found them rapidly dwindling in size and sacrificing their intellectual vitality to a preoccupation with the needs of church and state.[34] Paris, likewise, presented at this time a sad spectacle of fading greatness. The end of the French civil wars left her fallen in numbers and poorer, perhaps, than ever before.[35] Intent upon the preservation of her privileges and increasingly subject to a royal authority that, in the later years of the seventeenth century, seemed intent on keeping her docile and closed to any unsettling currents of thought, she subsided into intellectual sterility.[36]

Across the Channel, Oxford and Cambridge had continued to be absorbed in theological polemic until the middle of the seventeenth centu-

---

[31] *Nieuw Nederlandsch Biografisch Woordenboek*, ed. P. C. Molhuysen and P. J. Blok (Leiden: A. W. Sijthoff's Uitgevers-Maatschappij, 1911-37), Vol. VI, 127-41.

[32] d'Irsay, *Histoire des universités*, Vol. II, p. 85. For recent attempts to assess Boerhaave's significance, see *Boerhaave and His Time*, ed. G. A. Lindeboom (Leiden: E. J. Brill, 1970).

[33] Mrs. H. M. Vernon, *Italy from 1494 to 1790* (Cambridge: The University Press, 1909), p. 287. Ornstein, *The Rôle of the Scientific Societies*, pp. 217-9. Castiglioni, *A History of Medicine*, pp. 487 and 568.

[34] See Richard L. Kagan, "Universities in Castile, 1500-1700," *Past and Present*, no. 49 (Nov. 1970), pp. 44-71.

[35] Orest Ranum, *Paris in the Age of Absolutism: An Essay* (New York, etc.: John Wiley and Sons, c. 1968), pp. 8 and 21. Charles Jourdain, *Histoire de l'Université de Paris au XVIIe et au XVIIIe siècle* (Paris: L. Hachette et Cie, 1862-6), p. 32.

[36] Jourdain, *op. cit., passim.*

ry, when Oxford briefly flourished under the care of the Puritans.[37] Following the Restoration, however, both universities were gripped by a distrust of innovation and entered a period of diminishing numbers and diminishing influence in English intellectual life.[38] The five (or four) universities in seventeenth-century Scotland, meanwhile, remained, under the close surveillance of the Church, little more than small liberal arts colleges offering accessory instruction in theology.[39] In Germany, a rapid growth in the number of universities, both Catholic and Protestant, reflected less an enthusiasm for learning than the progressive political and confessional fragmentation of the Empire.[40] The faculties of theology dominated the German schools throughout much of the seventeenth century, and perhaps at no other time in their history was the concern for controlling education for the sake of orthodoxy so pronounced.[41] During the second half of the century, the respect the German universities commanded within Germany itself reached its nadir, and it was only in the last decade, in 1694, that the founding of the University of Halle by the Hohenzollern Elector of Brandenburg prepared the way for a revival of German higher education.[42] In the meantime, the University of Leiden, by virtue of her vigor and continuing influence in the society about her, stood out as a striking exception in what, otherwise, was a bleak chapter in the history of European universities.

The school at Leiden, of course, was the product of that very society that closely surrounded and sustained it, a society in which many of the most powerful forces reshaping European life and thought were most keenly felt. In the seventeenth century, the Dutch Republic was the home *par excellence* of commerce and capitalism, and the city of Leiden itself, second largest of the republic's rapidly growing towns, was one of

---

[37] Mullinger, *The University of Cambridge*, Vol. II, pp. 96 and 284. Morison, *The Founding of Harvard College*, pp. 58-9. Mallet, *History of the University of Oxford*, Vol. II, pp. 391-402 and 463.

[38] Hugh Kearney, *Scholars and Gentlemen: Universities and Society in Pre-Industrial Britain, 1500-1700* (London: Faber and Faber, 1970), pp. 141-57.

[39] H. M. Knox, *Two Hundred and Fifty Years of Scottish Education, 1696-1946*, (Edinburgh and London: Oliver and Boyd, 1953), pp. 14-7. Morison, *The Founding of Harvard College*, p. 128. Kearney, *Scholars and Gentlemen*, pp. 129 ff. and 154-6. Alexander Morgan, *Scottish University Studies* (London: Oxford University Press, 1933), p. 56.

[40] Friedrich Paulsen, *Geschichte des gelehrten Unterrichts* (3rd ed.; Leipzig: Veit und Comp., 1919-21), Vol. I, pp. 190-202 and 257-9. Ornstein, *The Rôle of the Scientific Societies*, pp. 226-7.

[41] Friedrich Paulsen, *The German Universities and University Study* (trans. Frank Thilly and William W. Elwang; New York: Charles Scribner's Sons, 1906), pp. 37-8.

[42] Paulsen, *Geschichte des gelehrten Unterrichts*, Vol. I, p. 524.

[43] Leiden, numbering about 45,000 inhabitants in 1622, had grown to at least 60,000 well before the end of the seventeenth century. By the eighteenth century,

Europe's greatest industrial centers.[43] Among her leading industries, moreover, was that of printing, vehicle for the continuing battle of ideas,[44] and the Dutch oligarchy sheltered a spirit of tolerance which, though far from complete, was advanced for its time. The diverse cross currents of the turbulent intellectual life of the century found their freest expression under this regime, and it was here that the ideology of liberty, despite the shortcomings of its realization and the special interests it cloaked, was most vigorously embraced. In such a milieu, how difficult it would have been for the University of Leiden – "trophy," one of her professors called her, "of the restored liberty in the entire commonwealth" [45] – to remain shuttered against the radical new developments in European thought, the most consequential, and the most disturbing, being the birth of modern science.

The ideas, attitudes and practices that together constituted the new natural science emerging in the seventeenth century soon, indeed, began to make their influence felt at Leiden, but here, as elsewhere, the new science would prove difficult to assimilate, for it challenged the knowledge, procedures and aspirations cultivated by centuries of academic learning. It was within the traditional discipline of "physics," or "natural philosophy," that the most direct confrontation took place between the new science and the inherited system of knowledge, and until new fields

however, the population of the city had begun to fall sharply, though she remained the republic's second largest city well into the new century. Amsterdam, of course, remained the largest, with about 100,000 inhabitants in 1622 and over 150,000 a little more than a century later. (J. A. Faber, *et. al.*, "Population Changes and Economic Developments in the Netherlands: A Historical Survey," *A. A. G. Bijdragen*, Vol. XII [1965], pp. 56-7. Roger Mols, *Introduction à la démographie historique des villes d'Europe du XIVe au XVIIIe siècle* [Louvain: J. Duculot, 1954-6], Vol. II, pp. 522-3.)

Charles Wilson also describes Leiden as the largest single industrial concentration in Europe by the mid-seventeenth century, surpassed as an industrial town only by Lyons. (*Economic History and the Historian: Collected Essays* [New York: Frederick A. Praeger, 1969], pp. 116 and 101).

[44] C. T. Smith, *An Historical Geography of Western Europe before 1800* (London and Harlow: Longmans, Green and Co. Ltd., 1967), p. 456.

[45] Joannes de Raey, *Clavis philosophiae naturalis seu introductio ad naturae contemplationem, Aristotelico-Cartesiana* (Lugd. Batavor.: Ex Officina Johannis et Danielis Elsevier, 1654), "Epistola dedicatoria."

[46] At the request of Prince Maurice, a chair for mathematical instruction in Dutch for engineers and military officers was indeed created and attached to the University of Leiden in 1600. The organization of the instruction, promisingly enough, was outlined by Simon Stevin (see *Bronnen*, Vol. I, pp. 389*-92*). From 1615 until 1679, the chair was occupied by three successive members of the Van Schooten family of mathematicians; it was apparently suppressed in 1681 but reestablished again at the turn of the century. (J. J. Woltjer, *De Leidse Universiteit in Verleden en Heden* [Universitaire Pers Leiden, 1965], pp. 28-9. *Bronnen*, Vol. IV, pp. 158-9 and 189.)

of study were recognized by the universities,[46] it was within the framework of this discipline – and, to a lesser degree, within that of medicine as well – that the new science had to find its way into the lecture halls. The faculty responsible for the discipline of physics, however, the faculty of philosophy, was not in a position to do justice to the importance of what was taking place. On the continent, it was a faculty that addressed itself primarily to the younger boys – the curators at Leiden had to anticipate students under fourteen years of age [47] – and the instruction it offered was in the process of sinking to the level of secondary education.[48] Traditionally a preparatory background for subsequent study in theology, medicine or law within the universities, there was no higher faculty of philosophy, and students generally completed their philosophy curriculum and acquired the Master of Arts degree, if they bothered to acquire it at all, before they were out of their teens. The tendency of the last centuries, indeed, had been to render the subject matter of philosophy more rapidly accessible to younger minds, and, despite the fact that physics was normally at the end of the philosophy curriculum, it could hardly provide a framework in which the development of the new science could be considered with sufficient sophistication.

Few degrees in philosophy were sought on the continent, for most students had, from the beginning, set their sights on one or another of the higher faculties.[49] In the Dutch universities, there were no entrance or candidacy exams and the students generally were free to determine their own course of study; as a consequence, students frequently left the philosophy curriculum unfinished as they rapidly moved on to the more advanced disciplines.[50] Initially, at least, candidates in theology at Leiden were required to delve into physics,[51] and medical students had always stressed physics in their philosophical orientation. Only the law student,

---

[47] *Bronnen,* Vol. I, p. 315*.
[48] Walter J. Ong, *Ramus, Method, and the Decay of Dialogue* (Cambridge, Mass.: Harvard University Press, 1958), pp. 136-7. It was otherwise in England, where the upper faculties had withered away, leaving the arts, or philosophy, curriculum predominant in university education. Here, however, the tutorial system also worked against a sophisticated treatment of individual fields. In the Italian universities, philosophical instruction was offered as a preliminary to medical study in particular. (Hastings Rashdall, *The Universities of Europe in the Middle Ages,* ed. F. M. Powicke and A. B. Emden [Oxford University Press, 1936]; Vol. I, p. 235; Vol. II, pp. 320-1; Vol. III, p. 151.)
[49] A notable exception would be the student body at the University of Paris, where all those enrolling in the higher faculties had first to have received a Master of Arts degree. (Roland Mousnier, *Paris au XVIIe siècle* (Paris: Centre de documentation universitaire, [1961]), p. 296.
[50] de Vrankrijker, *Vier Eeuwen,* pp. 71 and 60.
[51] *Ibid.,* p. 72.

perhaps, might feel free to ignore natural philosophy completely, yet it remained but one part of what, as a whole, was merely a preliminary and often neglected preparation for other fields of learning.

Throughout all of Europe, moreover, this preparatory instruction in philosophy had remained under the influence of Aristotle. He had ruled the schools – by consensus, however, rather than by decree – since the thirteenth century, and the schools, in turn, had dominated the intellectual life of Europe. The academic milieu, nonetheless, had then been a vital and questioning one; in the thirteenth and fourteenth centuries, anticipations of the seventeenth-century scientific revolution itself had appeared in the works of scholars at Oxford and Paris. But the fifteenth century had witnessed a decline in the creative energies in the universities, and the grip of philosophic tradition had become more rigid and unyielding. A reaction against the distortions and complexities of the Aristotelianism of the universities had come with the Renaissance, but without significantly diminishing the hegemony of Aristotle himself. The authority of a purified Aristotle, and of the classic past in general, was, if anything, enhanced. With the opening of the Reformation, a sweeping repudiation of Aristotle seemed at first to threaten, but his authority was soon reconfirmed with an even greater inflexibility than before. As the universities became more deeply involved in the combat between the churches, the growing hypersensitivity to doctrinal deviation made itself felt in the realm of philosophy as well. The unorthodox in philosophy was perceived all the more readily as a threat to orthodoxy in matters of faith and aroused those passions which a threat to the faith inevitably called forth.

Aristotle's monopoly in the schools, however, can easily be exaggerated. In physics, other classical authors, including, even if only to be refuted, Lucretius, the eloquent spokesman for the rival system of the atomists, were frequently cited. Moreover, as the foes of scholasticism were quick to point out, the natural philosophy of the schoolmen themselves embraced a multitude of contending interpretations, revisions, and elaborations of the Aristotelian texts that might have caused Aristotle no little distress. Nonetheless, the terminology of the peripatetic tradition, its basic concepts and problems, and the dialectical method it employed continued to determine the discipline of physics as taught in the universities.

Indeed, Aristotelianism in the schools rested on a deeper foundation than even the time-honored prestige of Aristotle or the anxiety of the churches. The peripatetic tradition corresponded to deeply ingrained patterns of perception and habits of thought shared by the great majority of European intellectuals. The new science demanded, in fact, the repudiation of a broad complex of traditional outlooks pertinent not only to the interpretation of nature, but to the idea of learning and knowledge as well. Nowhere did these outlooks have greater meaning than in the universities, and even those academics eager to espouse the new science would find it no easy task to discard the old.

Nonetheless, a succession of capable and conscientious professors at the University of Leiden responded to the challenge. In the seventeenth century, which Whitehead has dubbed "The Century of Genius," they trailed well behind the scientific advance; they were teachers first, philosophers second, and scientists, if at all, only last. Though seldom grasping the full meaning of the new science, however, they were increasingly sensitive to its influence and by the end of the century had committed the University of Leiden to its support. They prepared the way for their eighteenth-century successors who were to establish Leiden as one of Europe's most effective centers for the propagation of the methods and attitudes of modern science in its first centuries.

CHAPTER II

FRANCO BURGERSDIJCK:
LATE SCHOLASTICISM AT LEIDEN

Despite the prestige that the study of classical letters and history had soon acquired at the new academy, the disciplines of philosophy proper – logic, ethics and physics – initially suffered from consistent neglect.[1] The chairs designated for these subjects were frequently to be found vacant, and those who did occupy them were often drawn from other fields than philosophy. Only in the early seventeenth century does a greater concern for philosophic instruction become evident in the person of the Scot, Gilbertus Jacchaeus. Having come to Leiden as a student in theology in 1603, he joined the faculty of philosophy as professor of logic in 1605 and was two years later entrusted with the chair in ethics as well.[2] In 1614, however, he published his *Institutiones physicae,* the first work of any note on physics to issue from Leiden's faculty, and in 1617 he was given the responsibility for physics alone.

Among Jacchaeus' colleagues at this time was Willebrord Snellius, renowned mathematician and discoverer of the sine law of refraction. As professor of mathematics, Snellius appears to have begun offering a course in optics in the very year Jacchaeus assumed the professorship of physics.[3] Having himself carried out numerous experiments in optics, it is possible that before he died in 1626, Snellius provided the students at Leiden with a fleeting glimpse of the methods and mentality of the

---

[1] See Paul Dibon, *La Philosophie néerlandaise au siècle d'or,* Vol. I: *L'Enseignement philosophique dans les universités à l'époque pré-cartésienne, 1575-1650* (Paris, etc.: Elsevier Publishing Company, 1954); also Ferdinand Sassen, "Het oudste wijsgeerig Onderwijs te Leiden (1575-1619)," *Mededeelingen, Kon. Akademie van Wetenschappen, Amsterdam, Afdeeling Letterkunde,* new series, Vol. IV (1941).

[2] For a short biography of Jacchaeus, see the *Nieuw Nederlandsch Biografisch Woordenboek,* ed. P. C. Molhuysen and P. J. Blok (Leiden: A. W. Sijthoff, 1911-37), Vol. I, 1197-8.

[3] On Snellius, see *Ibid.,* Vol. VII, 1155-63, and J. A. Vollgraff, "Pierre de la Ramée (1515-1572) et Willebrord Snell van Royen (1580-1626)," *Janus,* XVIII (1913), pp. 595-625.

new science. But it was the *Institutiones* of the philosopher Jacchaeus, not the experimental investigations of the mathematician Snellius, that represented the physics taught in the universities of Europe, a science of nature that remained Aristotelian and scholastic, oriented not towards the investigation of nature but towards the verbal disputation that characterized the pedagogy of the schools.

The most influential representative of scholastic Aristotelianism in the republic, however, was Jacchaeus' student and successor at Leiden, Franco Burgersdijck.[4] The appointment of Burgersdijck to the faculty at Leiden followed as a consequence of the political and religious upheaval that shook the republic in 1618 and 1619, culminating in the Synod of Dort and the execution of Johan van Oldenbarnevelt. As the struggle between the House of Orange and the oligarchy of Holland made its grim debut, the rigidly orthodox party of the Dutch Reformed Church overcame the more latitudinarian party of the Arminians and then aspired to a domination of the intellectual life of the republic. Control of the universities would be vital, and the ministers directed their attention to the University of Leiden in particular, where the Arminian movement itself had originated. Their efforts to acquire a decisive voice in the administration of the university would ultimately end in frustration, for the new curators appointed during the crisis would prove no more sympathetic to ecclesiastical intervention than their predecessors, but, for the moment, the triumphant churchmen, allies of the Prince of Orange, were not to be resisted.

Although the character of the philosophic instruction at Leiden was not at issue, the faculty of philosophy, tainted by associations with the Arminian party, suffered most from the purge inflicted on the university. By the end of 1619, all three of the chairs in philosophy, each of which had been individually occupied at the beginning of the year, were vacant. Jacchaeus, the most fortunate of the purged professors, was restored to his chair in physics in 1623, but the empty chairs of both logic and ethics were quickly filled by a single new appointee, Burgersdijck, who was called from his post at the Academy of Saumur. Following the death of Jacchaeus in 1628, Burgersdijck exchanged his chair in ethics for that in physics, which he then continued to occupy until his own death in 1635. His manuals on physics, ethics, logic and metaphysics became not only the standard texts at Leiden but made him the most eminent of the Dutch Aristotelians far beyond the borders of the republic. His works on logic and metaphysics would remain recommended reading at the

[4] For biographical data on Burgersdijck, see the *Nieuw Nederlandsch Biografisch Woordenboek,* Vol. VII, 229-31.

English universities well into the eighteenth century, while, in his own time, his textbooks on physics were clearly among the most popular and influential of such introductory manuals.[5] His influence dominated philosophic instruction at Leiden until the crisis brought on by the challenge of Cartesianism.[6]

In 1622, six years before assuming the professorship of physics, Burgersdijck had published an *Idea philosophiae naturalis, sive methodus definitionum et controversiarum physicarum*;[7] a decade later followed a more elaborate *Collegium physicum*.[8] Both works testify to the pedagogical concern that marked so many of the academic publications in philosophy in the early seventeenth century.[9] The *Idea*, a "method of definitions and controversies pertaining to physics," reflected not only the disputative practices of pedagogical procedure in the universities but, as well, the preoccupation of late sixteenth and seventeenth-century academic philosophers with the organization and simplification of subject matter for the sake of more facile teaching. This organization and simplification they spoke of as their "method," and the method each evolved was, in their eyes, their most significant personal contribution to the discipline.[10] Burgersdijck wrote that he had published the *Idea* to demonstrate the order to be followed in the study of physics and to provide those students studying at home with continual theses to debate.[11] From the diffuse works – "well-nigh without method" – of recent commentators on Aristotle, he had selected disputations on physics and arranged them "according to that method which I believe best suited to the nature of this science." The prospective students were assured that if they followed this method, their studies could be greatly abbreviated.

This preoccupation with what was regarded as pedagogical efficiency, however, hardly encouraged a greater receptivity towards the innovations of the new science. Anything beyond a superficial survey – of either

---

[5] Nicholas Hans, *New Trends in Education in the Eighteenth Century* (London: Routledge and Kegan Paul Ltd., 1951), p. 51. Kearney, *Scholars and Gentlemen*, p. 164. Sister Mary Richard Reif, "Natural Philosophy in Some Early Seventeenth Century Scholastic Textbooks" (unpublished Ph.D. dissertation, Saint Louis University, 1962), p. 5.

[6] Dibon, *La Philosophie néerlandaise*, Vol. I, p. 94.

[7] The second edition (Lugd. Batavorum: Ex officina Bonavent. et Abrahami Elzevir, 1627) has been used in this study.

[8] Again, the second edition (Lugd. Batavorum: Ex officinâ Elziviriorum, 1642) will be referred to below.

[9] See Neal W. Gilbert, *Renaissance Concepts of Method* (New York: Columbia University Press, 1960), *passim*.

[10] Reif, "Natural Philosophy," pp. 269-70 and 279-81. Gilbert, *Renaissance Concepts, passim*.

[11] *Idea philosophiae naturalis*, "Philosophiae studiosis."

Aristotelianism or the new science, for that matter – was clearly precluded, but, further, the philosophical difficulties and anomalies unavoidably raised by the new science would surely be unwelcome where a neat systematization was so highly esteemed. The pedagogical method of the text-book writers was based on a prior systematic structure perceived within the subject matter itself.[12] When new ideas threatened the order and coherence of the philosophic system, those ideas necessarily threatened "method" in philosophic instruction as well.

Consisting of thirty-four disputations defended by Burgersdijck's students the year preceding publication,[13] the *Collegium physicum* reflects the general organization of Burgersdijck's course. After beginning with a consideration of physics as a discipline within the broader framework of philosophy, the *Collegium* proceeds to the three primary "principles" of peripatetic physics – matter, form and privation – and then to the "nature, efficient cause and end [in the sense of goal or purpose] of natural things." Following disputations deal with the nature of magnitude in natural objects, the nature of place and void, motion in general, specific types of motion, and time. Disputation ten concerns the heavens; eleven, the stars; and the following two, the elements. There follows a disputation on the origin and destruction of things, *"De ortu et interitu,"* then atmospheric phenomena, mixture and decomposition *("De mixtione et petredine"),* the constitution of mixtures, and the kinds of mixtures. With disputation twenty begins the treatment of organic nature, including the psychology of man, traditionally the most important part of scholastic natural philosophy.[14] The living thing and the essence of life are considered, then life and death, nutrition and growth, the generation of living things, sense and the senses, intelligence, intellect, and the will. All is finally concluded with a last disputation on "the world," *"De mundo."*

Scholastic physics was, therefore, a broad and inclusive discipline that embraced speculation pertaining to many of the modern sciences, from psychology to meteorology. Within such a comprehensive framework, only scattered doctrines were to be directly challenged by the innovations of the new science. Among those called into question, however, were

---

[12] See Reif on the manualists' discussion of method, "Natural Philosophy," pp. 267-81.

[13] Dibon, *La Philosophie néerlandaise*, Vol. I, pp. 44 and 96.

[14] Maurice de Wulf, *Scholasticism Old and New: An Introduction to Scholastic Philosophy, Medieval and Modern* (trans. P. Coffey; Dublin: M. H. Gill and Son, Ltd.; London: Longmans, Green and Co., 1910), p. 123. William T. Costello, *The Scholastic Curriculum at Early 17th Century Cambridge* (Cambridge, Mass.: Harvard University Press, 1958), p. 94. In the years around 1600, the soul was, in fact, the most popular subject for disputations in physics at Leiden (Dibon, *La Philosophie néerlandaise,* Vol. I, pp. 57-8).

doctrines fundamental to the scholastic philosopher's understanding of nature and to the internal cohesiveness of his physics.

"The world" had a very precise meaning for Burgersdijck; it was the "assemblage (*collectio*) of natural bodies," [15] and it was these "natural bodies" with which his physics dealt. The investigation of these natural bodies, however, began with questions that would have little meaning within the context of the new science. The natural bodies of Burgersdijck's physics were, in effect, the myriad corporeal identities that were perceived in nature, and the identity of each natural body was conceived as a basic fact that could be resolved into no simpler components or concepts. Nonetheless, the natural body itself, and hence its identity, was also necessarily transient and impermanent. According to the etymology of the word "physics," Burgersdijck pointed out, it was, indeed, a science of things that "once having arisen, afterwards perish." [16] A primary task of his physics, therefore, would be to provide an understanding of the processes through which the natural body as a specific identity came to be and then passed away, its "generation," that is, and "corruption."

"Generation" and "corruption" were the extreme processes of change with which natural philosophy could deal; beyond were only "creation" and "annihilation," in which a thing emerged from absolutely nothing and returned to absolutely nothing. The common definition of generation and corruption, said Burgersdijck, was Aristotle's: a change in which something is altered in its entirety and nothing in that thing remains perceptibly the same.[17] It was essential that it be understood that this was not a case of creation or annihilation, "For above all," Burgersdijck declared, "we are to avoid affirming that anything comes naturally into being from nothing or passes away into nothing." [18] At the same time, however, it was no less essential to the meaning of generation and corruption that what took place was a transition between being and non-being.[19]

The processes of generation and corruption were conceived in terms of the three principles of scholastic natural philosophy, and, conversely, it was the undeniable rise and destruction of natural things that, for Burgersdijck, confirmed the existence of the principles.[20] The best explanation of these principles, acknowledged Burgersdijck, was to be

[15] *Collegium physicum*, p. 343.
[16] *Ibid.*, p. 2.
[17] *Ibid.*, p. 136.
[18] *Ibid.*, p. 29.
[19] *Ibid.*, p. 137.
[20] *Ibid.*, p. 12.

sought within the discipline of metaphysics, but it was proper, nonetheless, that physics begin by demonstrating what they were.

The principle that best embodied the peripatetic perception of nature was form, for it was form that provided the identity of a natural body. Defined according to Aristotle, said Burgersdijck, form was "that by which a thing is what it is." [21] Joined to matter, the form of a thing established the nature of that thing and was responsible for its distinctiveness among the other natural bodies that composed the world.[22] Therefore, although there was only one matter that joined in the composition of all natural bodies, there were as many individual forms as there were individual bodies, "for since the distinction between natural things follows from their forms, there can be no less distinction between the forms than between the natural bodies themselves." [23]

Forms came into being and passed away simultaneously with their respective natural bodies, and it was the coming and going of form that provided the framework within which Burgersdijck attempted to reconcile the two limiting criteria of generation and corruption, that, on the one hand, generation and corruption were transitions between being and non-being and, on the other, that they were *not* transitions between being and nothing. Although forms came into being and then passed away, they were not, declared Burgersdijck, to be considered to have been utterly nothing before they arose nor after they perished.[24] "We say, then, with Aristotle that forms are not nothing before generation; rather, that which they are by actualization after generation is potentiality before," [25] and that which was potentiality, he explained only parenthetically, was not completely nothing.

Not potentiality, however, but privation was the third principle of Burgersdijck's physics, and, though similar in some respects, the two were not to be confused. Privation was the absence of a certain form in something otherwise fit to receive that form, and whereas potentiality could no more be destroyed than matter, within which potentiality was contained, privation passed away with the coming of the form.[26] Privation, as an absence of form, was itself "non-being," but a kind of non-being that was not nothing at all but was the lack of something specific,

---

[21] *Ibid.,* p. 25 ff.
[22] *Ibid.,* p. 345.
[23] *Ibid.,* p. 28. Nonetheless, the Aristotelian form established the identity of a thing essentially with respect to its "species."
[24] *Ibid.,* pp. 29-30.
[25] *Nos igitur cum Aristotele statuamus, formas ante generationem non esse nihil, sed id esse potentia, quod actu sunt post generationem. (Ibid.)*
[26] *Ibid.,* pp. 31-2.

such as non-man. Like potentiality, privation was, for Burgersdijck, something negative, but, nonetheless, it was something. The continuous alternation of privation and form within matter gave rise to the changing succession of natural bodies in the world, and, hence, the generation and corruption of natural bodies was, indeed, the coming and going of being, but it was not, as Burgersdijck understood it, a coming from or going back to nothing.

While forms, and therefore natural bodies, continually came and went, the principle of matter persisted throughout. It was the subject from which all natural bodies were composed and into which they ultimately resolved when they perished.[27] This matter was not some corporeal stuff, however, for corporeality was produced only when matter and form were joined.[28] Matter was in the first place a logical complement to form and privation, and it was through the logical investigation and manipulation of such concepts that Burgersdijck and his fellow scholastics hoped to acquire and convey an understanding of nature. The comprehension of the truths exposed through logic, however, was itself no easy task, and such late scholastics as Burgersdijck gave occasional evidence that they themselves were not completely happy with their own understanding of basic concepts they were teaching.

Burgersdijck's principle of matter was in fact Aristotle's "first matter," and the difficulty of understanding first matter was generally conceded in the scholastic manuals of the time.[29] According to Burgersdijck, it was best understood through Aristotle's analogy of the bronze statue: as the bronze is to the statue, so is first matter to the things which are composed of it, and just as the bronze is not "actually" (*actu*) the statue before the bronze accepts the form of the statue, so first matter is likewise only "potentially" (*potestate*), not "actually," the heavens, earth or whatever.[30] This matter was devoid of all forms but capable of receiving all forms and thus, explained Burgersdijck, was said to be "pure potentiality," a Thomistic formula for first matter that scholastic natural philosophers of the early seventeenth century generally rejected, in fact, as inadequate.[31] Burgersdijck affirmed, nonetheless, that the universal potentiality for all forms was the very essence of matter, that by which matter was what it was.

We are told explicitly that this matter is not physical body, for it lacks

[27] *Ibid.*, p. 14 ff.
[28] *Ibid.*, p. 26.
[29] Reif, "Natural Philosophy," pp. 105-6.
[30] *Collegium physicum,* pp. 20-1. "Act," *actus,* had a formal meaning in peripatetic philosophy; it was the perfected realization of what was previously potential.
[31] Reif, "Natural Philosophy," pp. 110-1.

any shape, sensible qualities or definite quantity (the quantity possessed by first matter was indeterminate quantity).[32] Nonetheless, Burgersdijck declared matter to have an independent existence of its own, though it did not "naturally" exist independently of form.[33] He explained that just as the bronze, though only potentially a statue, was itself still actual body, so matter likewise existed in itself.[34] If it were otherwise, he could not see how matter entered into the constitution of bodies nor what was to prevent its being called pure nothing, a sentiment shared by not a few seventeenth-century anti-scholastics as well.

Burgersdijck likewise granted an independent reality to form, even more reality, indeed, than he had attributed to matter, and in adhering thus to a scholastic tradition that, in contrast to Aquinas, endowed matter and form with their own reality, he again exemplified the scholastic pedagogues of his time.[35] It has been observed that, in defending the independent reality of matter and form, the scholastic textbook authors in the early seventeenth century appear to have been attempting to conceive these concepts of logical analysis as more imaginable components of being.[36] Could this unconscious effort have reflected an uneasiness alluded to above, a longing for a more satisfying understanding of the scholastic principles? At Leiden, the issue of the principles of physics would loom large in the subsequent clash between philosophies, and perhaps we may already detect among the late scholastics themselves a malaise from which, for many of their heirs, only the mechanical philosophy would offer relief.

As taught by Burgersdijck, the principle of form was in effect the "substantial form" so much disparaged by the "moderns" of the seventeenth century. It represented, indeed, an orientation in scholastic Aristotelianism that emphasized what was least compatible with the perspective of the evolving new science, for it embodied the idea of identity as a basic concept and reflected a concern with the most radical of the Aristotelian natural processes, the generation and corruption of "substance." Substance was the most fundamental kind of being; it was able to exist by itself and thus provided the foundation in which lesser kinds of being, "accidents," could occur. But peripatetic substance was also necessarily a specific identity, and though such accidents as quantity, color, shape and place provided the sensible appearance of a thing, the

---

[32] *Collegium physicum*, pp. 20-3.
[33] *Ibid.*, p. 139.
[34] *Ibid.*, p. 21.
[35] *Ibid.*, p. 26. Reif, "Natural Philosophy," pp. 44-5 and 128-30.
[36] Reif, *op. cit.*, p. 128.

identity they thus revealed was determined by the substance in which they resided.

In asserting the independent reality of form and matter, Burgersdijck, like other late scholastics, had declared form itself to be a "true substance" and granted matter "substantial existence," [37] but the most perfect and complete substance in physics was the natural body itself, which was therefore the host, as it were, of the accidents that constituted its appearance. Hence, in treating the principle of form primarily with regard to the identity of natural bodies, Burgersdijck was treating it as the form of substance, substantial form.[38] Conversely, when he chose to consider the processes of generation and corruption only in so far as they applied to substance, he was persisting in his preoccupation with the coming and going of natural bodies in nature.[39]

Presented largely in the guise of first matter and substantial form, the principles of Burgersdijck's natural philosophy were the Aristotelian principles in an extreme and simplistic (and distorted) rendering. In part the consequence, no doubt, of his commitment to the expeditious teaching of the young, this oversimplification, disregarding the accidents, stressed precisely those aspects of the peripatetic principles that were most alien to the new science. Burgersdijck's fellow scholastics in the early seventeenth century, it is true, did not all concur in such an extreme representation of the principles of Aristotelian physics.[40] But many did, and even those who did not still built their natural philosophy around the basic idea of the generation and corruption of substance,[41] processes that were largely meaningless within those branches of the new science that were most significant in reshaping scientific thought in the seventeenth century. For those critical areas of scientific activity, neither the coming and going of substantial being nor the natural body as a distinctive identity were concepts of much value. As a consequence, the growing influence of the new science doubtless contributed to the increasing disillusionment with the scholastic principles, often rejected now as mere word play without meaning.

But Aristotelian physics was by no means limited, after all, to the consideration of the vicissitudes of substance, and even Burgersdijck, despite the bias in his treatment of the Aristotelian principles, elsewhere also stressed the paramount importance of change pertaining to accidents,

---

[37] *Collegium physicum*, pp. 26 and 21. Reif, *op. cit.*, see pp. 119-20 and 126 ff.
[38] *Collegium physicum*, see pp. 25-6.
[39] *Ibid.*, p. 136.
[40] Reif, "Natural Philosophy," see pp. 82-5.
[41] *Ibid.*, p. 103.

the importance, that is, of motion. The subject of motion, indeed, had long been central to natural philosophy. Aristotle had considered what was called "natural motion" the hallmark of all that fell within the province of physics, for all that physics dealt with had to be "natural," and that meant, precisely, that everything dealt with possessed an internal source of motion. Aristotle, it is true, had also held the generation and corruption of substance to be a kind of motion, but Burgersdijck, though unwilling to accept any type of generation and corruption as motion, still followed Aristotle nonetheless in declaring that the physicist studied natural bodies in so far as they had within them a principle of motion, "in so far, that is, as they are natural." [42] Without motion, he asserted, all causes and principles in natural philosophy would be abolished.[43] The scholastic manualists of the early seventeenth century in general attached a similar importance to the subject of motion,[44] and so likewise did the builders of the new science. Sir Isaac Newton, the supreme representative of the new science of the seventeenth century, would also affirm that "the whole burden of philosophy seems to consist in this – from the phenomena of motions to investigate the forces of nature, and then from these forces to demonstrate the other phenomena." [45] But the Aristotelian and Newtonian understanding of motion were literally worlds apart, and it was in its reconceptualization of motion, not in its indifference to the generation and corruption of substance, that the new science posed its most consequential challenge to the traditional philosophy of nature.

The doctrine of motion taught by Burgersdijck remained basically Aristotelian, and it conformed to the patterns of thought that had also shaped the peripatetic first principles. Burgersdijck explicitly distinguished motion from generation and corruption, for the former was gradual (*successiva*) while the latter were instantaneous (*momentaneae*).[46] Nonetheless, though the generation and corruption of accidents were neither motion nor parts of motion, they were the beginning and end of motion – just like the two extreme points of a line, explained Burgersdijck, which were neither the line nor parts of the line.[47] Motion, moreover, was understood in terms of a formula that echoed the explanation of generation: defined according to Aristotle, taught Burgersdijck, motion

---

[42] *Collegium physicum*, pp. 76 and 4.
[43] *Ibid.*, p. 16.
[44] Reif, "Natural Philosophy," pp. 213 and 215.
[45] *Sir Isaac Newton's Mathematical Principles of Natural Philosophy and His System of the World*, trans. Andrew Motte (1729), rev. by Florian Cajori (Berkeley, Calif.: University of California Press, 1946), pp. XVII and XVIII.
[46] *Collegium physicum*, p. 76.
[47] *Ibid.*, p. 78.

was the actualization of that which was potential to the extent that it was potential; motion, therefore, was also an advance from potentiality to actualization, from privation to form (though now the form of an accident).[48]

That this motion, when it was "natural," derived from a cause that was internal to the moving body was a tenet that was not open to debate. Likewise, that this internal source of motion was nothing less than the very "nature" of the body was also beyond question; to Burgersdijck and the other Aristotelians, it was a fact that was obvious from immediate experience.[49] "Nothing is better known in natural philosophy," asserted Burgersdijck, "than the fact that there is motion that arises from an internal principle, which is nothing other than to be nature [in a thing]." [50] To ask the natural philosopher to demonstrate the existence of that internal principle was utterly absurd, "for an infinite number of things are perceived that, since they possess that principle within themselves and by virtue of it continue to endure, are called natural." [51] Whatever lacked such an internal principle of motion was not "natural" and was to be excluded, therefore, from the realm of physics.[52]

Specifically so excluded by Burgersdijck for being "non-natural" were both the products of human craft (*res artificiosae*) and things mathematical,[53] both of which, however, were of fundamental importance to the new science. Like physics (and metaphysics), mathematics was one of the three "theoretical" sciences of scholastic philosophy, and Burgersdijck, exemplifying the peripatetic tradition, emphasized the distinctiveness of each of these sciences.[54] He recognized magnitude as an accident which natural bodies could not be without [55] and devoted a separate disputation to its consideration. Nonetheless, true to his proscription, Burgersdijck's treatment of quantity in natural bodies introduced no mathematics. As for the mathematicians, they were concerned with abstract mathematics, observed Burgersdijck, and when they dealt with natural bodies, as they did in astronomy and geography, they did so only with respect to the quantity of these bodies, not with respect to their

[48] *Ibid.*, pp. 63 and 73.
[49] Reif, "Natural Philosophy," pp. 145-6.
[50] *Collegium physicum*, p. 35.
[51] *Ibid.*, p. 34.
[52] *Ibid.*, pp. 37-8 and 4-6.
[53] *Ibid.*, p. 4.
[54] *Ibid.*, pp. 9-10. In addition to the theoretical sciences were the practical sciences, including, according to Burgersdijck, ethics, politics and economics (*philosophia oeconomica*). Logic, said Burgersdijck, should not be considered a part of philosophy but, rather, an instrument of philosophy. (*Ibid.*, pp. 1-2.)
[55] *Ibid.*, pp. 42-3.

being natural.⁵⁶ To the architects of the new science, however, it was precisely through the measurement and mathematical manipulation of quantity that nature was to be most effectively grasped, and their supreme achievements followed from their having rendered motion itself, the scholastic criterion for what was natural, mathematical.

The Aristotelians recognized other motion than "natural" motion, however: there was as well motion that was alien to nature, and, hence, was not "natural" but "violent." Burgersdijck explained violent motion as motion caused by an external principle resisted, or at least not abetted, by the internal principle.⁵⁷ Within the realm of violent motion was found the difficulty in Aristotelian physics, the problem of projectiles, that, in retrospect, most clearly anticipated the radical change in the understanding of motion that was to take place in the seventeenth century. Why did a stone which had been thrown continue its flight, a motion alien to its nature, after it had left the hand that threw it? "Since the mover must always be joined to the movable body," said Burgersdijck, "we seek here the cause by which the motion of projectiles is continued." ⁵⁸ In finding that cause to be a force or quality "impressed" on bodies, Burgersdijck himself adhered to the doctrine of an "impetus" that, after having been imparted to the projectile by the hand or instrument that threw it, gradually dissipated in the course of flight. Elaborated by fourteenth-century scholastics, the concept of impetus had been quietly incorporated into the body of peripatetic physics during the following centuries, but in the seventeenth century, the new science would find a revolutionary new solution to the problem of projectiles in the concept of inertia. As motion became inertial, however, the Aristotelian distinction between motion that was natural and motion that was violent had itself to be rejected.

Within the fuller context of Aristotle's own teachings, there was already another basic tenet that was not easily reconciled with the all-important idea of natural motion; as cited, and only briefly, by Burgersdijck, it declared that "nothing is moved by itself, but whatever is moved is moved by another." ⁵⁹ At first sight, this was an idea far more congenial, so it seemed, to the mechanical philosophy that was soon to challenge the Aristotelianism of the schools, and one of Burgersdijck's later successors at Leiden would stress this axiom in attempting to depict Aristotle as a proto-Cartesian. It was not, however, the internality as

⁵⁶ *Ibid.*, p. 5.
⁵⁷ *Ibid.*, p. 82.
⁵⁸ *Ibid.*, pp. 84-5.
⁵⁹ *Ibid.*, p. 66. See Friedrich Solmsen, *Aristotle's System of the Physical World* (Ithaca, N.Y.: Cornell University Press, 1960), pp. 94 ff. and 232-3.

such of Aristotelian natural motion that was antithetical to the new conception of motion being formulated; rather, it was its identification with the dynamics of self-realizing identities, natural bodies, and, hence, with a multitude of specific goals.

For Burgersdijck, all motion, to be motion, had to have both a *terminus a quo* and a *terminus ad quem*, the former being that which through motion was left behind, the latter, that which was reached or acquired.[60] The *terminus a quo*, conforming to a now familiar formula, was always the privation of the *terminus ad quem*, for motion was always the progress from privation to form (or from one form, destroyed through motion, to another contrary form). With the attainment of its *terminus ad quem*, motion came to an end, and since the state of rest now achieved followed no less from the nature of a body than had its prior motion, Aristotle had spoken of nature as the principle of not only motion but rest as well.[61] "And this quiet," wrote Burgersdijck, "follows actively from the same principle as does motion."

In fact, there were four components necessary in any motion – the *movens*, which caused the motion, the *mobile*, which was moved,[62] and the two *termini* – but it was the *terminus ad quem*, the fully realized form, that was the most complete expression of any motion. So much a part of the essence of motion was the *terminus ad quem*, said Burgersdijck, that motion could neither exist nor be understood without it; in some sense, indeed, it was identical with motion: "Motion is nothing other than the *terminus* or form which is produced...."[63] The concept of the *terminus ad quem* embodied the Aristotelian perception of all natural processes as teleological, the teleological character of nature having been, to Aristotle, the very foundation of her inherent order. Burgersdijck and his fellow scholastics in the early seventeenth century vigorously reaffirmed this understanding of nature,[64] and it is apparent from Burgersdijck's emphasis on the *terminus ad quem* that nothing was more important to his understanding of motion than his conviction that it was directed towards the achievement of some end. The conviction that motion was progress towards some specific goal lingered tenaciously throughout the century and proved a stubborn obstacle to the understanding and acceptance of a new motion that had, in the meantime, become inertial and relative.

[60] *Collegium physicum*, pp. 66 and 73.
[61] *Ibid.*, pp. 39-40 and 35.
[62] The *movens* and *mobile* preserved the idea that everything moved was moved by another.
[63] *Ibid.*, pp. 72-3.
[64] Reif, "Natural Philosophy," p. 200.

Aristotelian "motion" was also conceived as encompassing all manifestations of change, including, even when generation and corruption were excluded, change in quantity and quality as well as change in place. Indeed, it was qualitative change, "alteration," that provided Burgersdijck and his immediate successors at Leiden with their most frequent examples of motion, usually the heating or cooling of water. Like Aristotle himself, however, Burgersdijck and other scholastics granted a priority to the motion of changing place, "local motion," [65] and the depiction of all other types of change in terms of local motion alone would prove, at Leiden, one of the easier steps towards accommodating the new science. But even local motion was conceived by the scholastics as a passage from privation to form within the natural body. As Burgersdijck explained it:

Local motion does not strive towards place... but towards "where," or, rather, so to speak, towards "whereness" (*ubietatem*), towards that form or mode of existence, that is, which a body acquires from place. Every motion is the actualization of that which is movable, and in such a way that [the motion] is in the movable thing even as in a subject [that undergoes change]. Yet motion is not distinguished in reality from the uncompleted *terminus,* or in so far, that is, as [the *terminus*] has progressed in becoming, as they say. The *terminus* must be in the movable thing, therefore, as it would be in a subject. And since place is not in the thing which has been located there as it would be in a subject (it is, to be sure, an external adjunct), it follows that the *terminus* of local motion is not place but that, rather, which a body acquires when it enters a place.[66]

As a change that took place essentially *within* the body moved, even local motion was relevant only to the individual development of each natural body alone.

Nonetheless, this local motion so conceived also maintained the very structure of the world as an ordered whole, and this it accomplished through the four elements of the sublunar world and the fifth substance of the heavens. The local motion of inanimate nature in Aristotelian physics was either circular, the natural motion of the heavenly substance, or rectilinear, the natural motion of the elements that composed all bodies below the moon. Burgersdijck explained that the circular motion of the heavens was the only motion with no *terminus ad quem* at which it would come to rest; rather, it turned unceasingly about the center of the universe.[67] The rectilinear motion proper to the elements, meanwhile, moved either towards the center of the universe or away from it, though

[65] *Ibid.,* p. 212.
[66] *Collegium physicum,* pp. 80-1.
[67] *Ibid.,* pp. 73, 40 and 82-5.

never passing into the region of the moon or beyond. This dichotomy of rectilinear and circular motion was bound inextricably to the structure of the Aristotelian cosmos, divided into the fundamentally different regions of the heavens above and the terrestrial world at the center.

When Burgersdijck took up the structure of the cosmos, he reasoned backwards from the duality he had affirmed in local motion, arguing that since only one motion could be natural to an individual body, it followed that there were two types of bodies, the heavenly, whose natural motion was circular, and the elements, whose natural motion was rectilinear.[68] Aristotle had considered circular motion the most perfect motion, eternal and, hence, compatible with the divinity of the heavens, and Burgersdijck also echoed Aristotle in asserting that the "perfection" of the heavens argued their spherical shape:

> For since the heavens are a most perfect, most simple and most capacious (since everything is embraced within its periphery) body, it was appropriate to its nature that it have the most perfect, most simple and most capacious figure. None, moreover, is more perfect, more simple or more capacious than the spherical or orbicular, and this, therefore, is to be considered the most appropriate figure for the heavens.[69]

The inner surface of the heavens fit snugly around the elements – fire, air, water, and earth – which, in turn, were wrapped in successive spherical layers about the center of the cosmos, where, finally, the globe of the element earth resided.[70] The four elements, which could be resolved into no further constituents, were themselves the constituents of all sublunar bodies.[71] But the elements pertained less to chemical explanations than to the explanation of sublunar motion and cosmic order. They performed their primary function by virtue of the arrangement of their respective spheres, which were their respective natural places, where alone they could achieve a state of natural rest. Fire, tending to rise in all the other regions of the sublunar world, was located by its nature immediately below the heavens, while earth, which tended downwards in all other regions, occupied the very center of the world. Between them were air and water, the former immediately below the sphere of fire, the latter between the spheres of air and earth. Removed from their natural places by the disturbing influence of the heavenly motions and mixed together in the composition of natural bodies, the individual elements continually strove either upwards or downwards towards their proper spheres. When

---

[68] *Ibid.*, p. 96.
[69] *Ibid.*, p. 102.
[70] *Ibid.*, pp. 106 and 102-3.
[71] *Ibid.*, p. 116 ff.

they reached these spheres, their natural motion ceased, and they tended neither up nor down nor pressed one against the other.

The fundamental importance of the internal *terminus ad quem* of natural motion is all the more apparent; in addition to being the source of motion in sublunar bodies, the most immediate concern of physics, it was as well the very foundation of cosmic order.

> The elements, therefore, are moved naturally in rectilinear motion, and to this end, that they come to rest in the places appropriate to their natures and thereby set the universe in order.... Since this motion is truly natural to the elements, its principle must indeed be nature. Nature, however, is the principle of motion in that in which it is; consequently, the elements are moved not by an external principle, but by their own distinctive forms.[72]

Conversely, however, the concept of a cosmos so ordered by natural motion depended in turn on the implicit assumption that there could be "natural places," ineradicable differentiations underlying world space, differentiations, moreover, that could be somehow recognized by otherwise insensible natural bodies. This necessary association of a differentiated space with peripatetic local motion was to be challenged on all counts by the developed philosophic assumptions of seventeenth-century science. In the meantime, however, natural motion provided a common and essential understanding of both poles of the scholastic world, the stable and ordered cosmos, on the one hand, and the independent and ever-changing natural body, on the other.

Conceived more broadly as an actualization of potentiality, peripatetic natural motion also attested to the consistency of thought and conceptualization that continued to link scholastic physics to the formidable system of Aristotelian metaphysical speculation as well. Indeed, in so far as peripatetic motion was a quasi-metaphysical process of becoming, it was, like the peripatetic principles, inaccessible to the new science, for which it remained more alien than wrong. But as specifically embodied or exemplified in the organization and motions of the cosmos, there the scholastic doctrine of motion was exposed to direct contradiction. When the scholastic physicist turned to the more descriptive and particular treatment of the cosmos, an increasing number of specific points of conflict testified, ironically, that he and the early representatives of the new science were now, at last, thinking in similar terms and asking similar questions. It was in this realm of scholastic physics, therefore, that the new science made its first inroads.

In the sixteenth and early seventeenth centuries, the traditional as-

---

[72] *Ibid.*, p. 120.

tronomy had been called into question by new theories and dramatic new observations, and Burgersdijck acknowledged many of the problems that had been raised. He was entering no new paths, however, when he denied the physical reality of the geometrical devices of the Ptolemaic tradition, the epicycles, equants and eccentric circles, nor when he rejected as an "unadulterated fiction," at least with respect to the planets, the Aristotelian crystalline spheres, "that vaulting of solid orbs." [73] With respect to the fixed stars, indeed, he confessed that he could not explain their motion apart from some such solid orb, and asserted that they were either totally immobile, "in accordance with Copernicus," or, if they were moved, "which is the more widely accepted opinion," were moved by the motion of a stellar sphere. If it were otherwise, why, he asked, would the stars move all together in that same identical motion and not freely and independently like the planets, which moved most probably through a liquid heavens, "as birds move through the air or fishes through the sea"?

When Burgersdijck dealt with the incorruptibility of the heavens, however, the impact of recent astronomical observations became more immediately apparent. Aristotle had declared that the heavens were not only in perpetual motion but were perfect and incorruptible as well, and the heavens in scholastic physics remained a unique substance, immune to the imperfections, the changes, the generation and corruption that marked the region below the moon. Between 1572 and 1574, however, a new star had appeared and then passed away in the constellation of Cassiopeia, and the calculations of Tycho Brahe had confirmed with unprecedented accuracy that this star, as well as a succession of comets in the following decades, were located in that celestial region where no mutation was ever to occur. Other new stars had appeared as well in 1601 and 1604, and near the end of that decade, Galileo had first observed the rough and jagged surface of the moon through his telescope and, shortly after, the spots that moved across the face of the sun. The cosmos, divided into a perfect, immutable heavens and a disturbed and changing lower world, was appearing less divided.

Writing in the third and fourth decades of the century, Burgersdijck acknowledged to his students that "the certain demonstrations of the astronomers have recently made it impossible to deny any longer the birth of new stars and comets in the heavens," but he endeavored at the same time to diminish the significance of the awkward new observations and continued to maintain that the heavens were still fundamentally distinct from the elements and free of the corruption and generation that

[73] *Ibid.*, pp. 111-3.

distinguished the sublunar world.⁷⁴ He interpreted the appearance and disappearance of the new stars as but the condensation and rarefaction of the celestial substance, which remained, itself, essentially unchanged throughout.⁷⁵ The stars differed from the rest of the celestial substance only in their thickness, and though he granted them a form, which he said was light, it was a form that rendered them stars but not natural bodies, hardly a comfortable solution. Some parts of the heavens had been dense since the creation of the world, and these were the ordinary, imperishable stars, but others had condensed and then rarefied away again in subsequent ages, and these were the new stars and comets.⁷⁶ He offered as examples the new star of 1572 in Cassiopeia and the comet of 1618 and acknowledged "many others that would take too long to enumerate." ⁷⁷

Burgersdijck observed as well that all the parts of the heavens were not, after all, equally pure. This was apparent from the differences in the light of the eternal stars and from the spots upon the moon. It was probable that the substance of the moon was not so pure as the substance of the sun and stars, and the spots seemed to indicate that some parts of the moon itself were less pure than others. With Galileo's observations of the sunspots perhaps in mind, though unmentioned, Burgersdijck added that similar differences doubtless existed among the parts of glowing celestial substance as well.

The most striking facet of Burgersdijck's response to the new astronomy, however, was his hesitation with regard to Copernicus. "It is no easy matter," he admitted, "to resolve whether the hypothesis of Ptolemy is more true, declaring only the earth to be immobile, or that of Copernicus, asserting that the earth is moved with an annual and diurnal motion, while the fixed stars and sun together remain at rest." ⁷⁸ He acknowledged strong arguments on both sides and cited for his students that which he found most compelling in favor of Copernicus, taken from the *Utranometria* (1631) of his fellow Dutchman, Philippus van Lansbergen: if the diurnal motion were in fact attributed to the stars rather than to the earth, Burgersdijck declared, Saturn, in the space of about five minutes (*unico momento sive secundo horae scrupulo*), would have to traverse more than nine hundred German miles and the fixed stars, more than 643,000. This, "no one will easily concede to be possible," and if the earth, then,

---

⁷⁴ *Ibid.*, pp. 109-10 and 97 ff.
⁷⁵ *Ibid.*, pp. 106-7.
⁷⁶ *Ibid.*, p. 109 ff.
⁷⁷ *Ibid.*, p. 110.
⁷⁸ *Ibid.*, p. 113.

is granted a diurnal motion, an annual motion is much easier to believe as well.

Passed over in silence, however, was the fact that Copernicanism would overturn the whole cosmic physics which Burgersdijck had been teaching, the concentric order of the elements in the middle of the cosmos, their responsibility for natural motion in the sublunar world, the fundamental distinction between the earth and the celestial planets, and the eternal motion of the celestial substance. These consequences would seem much too obvious to have been overlooked; rather, they were far too shattering to the system of scholastic physics to be openly granted the same possibility that had been extended to Copernicanism as astronomical theory alone.

The Aristotelian cosmos was a compelling conception, wherein the most common observations of the universe had been organized into a simple and symmetric order revolving, as seemed only right, about man at the center, for whose sake, said Burgersdijck, the whole world could be said to have been put together.[79] The understanding of the cosmos was further interwoven with a highly developed body of thought seeking a deeper reality in nature and physical being. Such a sweeping systematization of philosophic thought and intuition was not lightly to be let slip, despite occasional flaws within its fabric or the incompatibility of certain scattered phenomena. It was a system within the framework of which European scholars had for centuries honed and tested their wits and for which the new science as yet offered no alternative of comparable breadth, order, or cohesiveness. As a demonstration of consistency and coherence, system itself was a weighty and, for many, necessary support to claims of truth for any body of philosophic thought, and within the academic community it had an added pedagogic value as well, for to surrender system in philosophy was to surrender the basis of "method." More broadly, to forsake system suggested a loss of purposefulness in teaching; the knowledge imparted to the young would appear undigested, its significance uncertain, its content aimless and arbitrary. System, and especially in the schools, had a value of its own that would preserve it well after new ideas and observations had begun to undermine the persuasiveness of the philosophy itself.

During Burgersdijck's years at Leiden, the challenge to scholastic physics was just beginning. The new astronomical theories and discoveries were but the preliminaries to a far more sweeping assault. In 1632, the year in which Burgersdijck's *Collegium physicum* had first appeared, Galileo's *Dialogo* on the alternative world systems, the Ptole-

[79] *Ibid.*, p. 351.

maic-Aristotelian and the Copernican, had also appeared at Florence. Five years later, in 1637, Descartes' *Discours de la méthode* was published by presses in Leiden itself, as was, in the following year, Galileo's *Discorsi e dimostrazioni matematiche* pertaining to "two new sciences." In 1641, Descartes' *Meditationes de prima philosophia* were printed at Paris, and in 1644, in Amsterdam, his *Principia philosophiae*. With the publication of the *Principia,* indeed, appeared a comprehensive new system of physical nature that was to emerge as the greatest rival to Aristotelianism within the universities.

CHAPTER III

TUMULT OVER CARTESIANISM

The philosophy of nature which the European reading public discovered in Descartes' *Principia philosophiae* depicted a new universe as dramatic in its sweep as it was compelling in the vividness of its imagery. Embracing many of the recent scientific discoveries and innovations, it shared as well in the aura of excitement that surrounded the new science, for which, indeed, Descartes had hoped to lay a secure philosophic foundation. The continuing progress of the new science would ultimately leave this new philosophy far behind, but during the remainder of the seventeenth century, that progress would take place largely against the background of the Cartesian world picture.

The outlines of Cartesian physics were known to Dutch intellectuals well before the publication of the *Principia*,[1] for it was in the Dutch Republic that Cartesianism sank its first roots and from whence it spread to the rest of Europe. In 1629, so as to be able to devote himself more completely to his philosophical efforts, Descartes had retired from the social entanglements of his native France to the United Provinces, where he made his residence until his departure for the Swedish court in 1649. An early fruit of his sojourn in the United Provinces had been the completion of his system of nature, and, though publication was forestalled for a decade by his alarm over the condemnation of Galileo in 1633, friends and sympathizers in Dutch academic circles became acquainted with Descartes' work.

Eventually penetrating Dutch society in a number of different guises, Cartesianism was to become an issue of heated controversy, for the theological implications attributed to Descartes' writings became the crux of the public reaction, hopelessly mixing the issue of Cartesianism with the bitter quarrels between theologians that plagued the intellectual life of the republic throughout the seventeenth century. Orthodox theologians looked with horror upon Descartes' method of universal doubt,

---

[1] Dibon, *La Philosophie néerlandaise,* Vol. I, 202.

which raised before them the specter of scepticism and unbelief.² Cartesian rationalism, in their eyes, threatened to overturn the always precarious relationship established between reason and revelation and to undermine the authority of the Scripture; they could not allow reason to exercise in philosophy an independence of theological surveillance.³ Against the background of the internal tensions that had rent the Calvinist movement itself, the Cartesian assertion of free will unavoidably evoked hostile reactions as well.⁴ Nor was Descartes' physics free of the theological polemics, and particularly offensive to the traditionalist theologians were Descartes' disdain for the substantial form, with which the soul had been identified, his implicit assertion of a moving earth in contradiction to explicit passages of the Bible, and his casual disregard for the Biblical "Genesis" in envisioning a Creation of his own.⁵

Nonetheless, despite the intense hostility, it was within the Dutch Republic that Cartesian physics won its first adherents, and, despite the entrenched position of the defenders of traditionalism and orthodoxy on the faculties, these adherents were chiefly to be found within the universities.⁶ In the republic, the universities quickly emerged, indeed, as the primary strongholds of the disciples of Descartes, and, unavoidably, it was within the schools that the first and major confrontations took place between the representatives of the old and the new philosophies.⁷

Descartes himself had numerous friends at Leiden; he often resided in or near the city and in 1630, the year after he had arrived in the republic, had reported to Burgersdijck himself, then rector, to enroll in the university.⁸ But it was rather at the newly-founded University of Utrecht, raised to the status of a university by the States of Utrecht in 1636, that the first of the clashes over Cartesianism broke out. The principals involved were Henri de Roy, better known as Regius, who had been ap-

---

² Josef Bohatec, *Die cartesianische Scholastik in der Philosophie und reformierten Dogmatik des 17. Jahrhunderts* (Leipzig: A. Deichert, 1912), pp. 22-4.

³ *Ibid.* Jan Anthony Cramer, *Abraham Heidanus en zijn Cartesianisme* (Utrecht: J. van Druten, 1889), p. 67. Ferdinand Sassen, *Geschiedenis van de Wijsbegeerte in Nederland tot het Einde der negentiende Eeuw* (Amsterdam en Brussel: Elsevier, 1959), pp. 142-3.

⁴ Cornelia Serrurier, *Descartes, l'homme et le penseur* (Paris: Presses Universitaires de France; Amsterdam: Éditions Françaises d'Amsterdam: [1951]),pp. 180-1.

⁵ Sassen, *Geschiedenis van de Wijsbegeerte in Nederland,* p. 143. Paul Mouy, *Le Développement de la physique cartésienne, 1646-1712* (Paris: J. Vrin, 1934), p. 17.

⁶ Mouy, *op. cit.,* pp. 13-4.

⁷ C. Louise Thijssen-Schoute, "Le Cartésianisme aux Pays-Bas," in *Descartes et le cartésianisme hollandais,* E. J. Dijksterhuis, *et al.* (Paris: Presses Universitaires de France; Amsterdam: Éditions Françaises d'Amsterdam; 1950), pp. 218 and 259.

⁸ Schotel, *De Academie te Leiden,* p. 154.

pointed professor of theoretical medicine and botany in 1638, and Gisbertus Voetius, professor of theology and, in 1641, rector of the university.[9] In the state crisis of 1618 and 1619, Voetius had acquired considerable stature as a leader of the victorious orthodox party, and now, two decades later, he set himself resolutely against the growing influence of Descartes. Through Regius, a young and enthusiastic disciple of the French philosopher, Voetius hoped to strike at Descartes himself. The conflict, which attracted much attention throughout the republic, began in 1639 with a disputation on atheism defended under Voetius' supervision. Certain theses clearly alluded to the Cartesian philosophy, and Regius took up Descartes' defence with a fervor that Descartes himself tried to temper. In 1642, after the academic senate at Utrecht had denounced recent disputations over which he himself had presided, Regius was instructed to limit himself henceforth to his proper domain of medicine (though Descartes' partial espousal of Harvey's theory of the circulation of the blood had also been an important point of contention). The statutes that were drawn up for the university in the following two years specifically restricted all philosophic instruction to the philosophy of Aristotle; no proponents of "absurd, paradoxical or novel doctrines" were to be tolerated, and philosophy was to be taught "so that the teachings of the Church are not exposed to ridicule and those who aspire to study in the higher faculties are not rendered useless," a threat directed as much, perhaps, to the aspiring students as to the members of the faculty.[10] Cartesianism continued to survive at Utrecht – in 1646 Regius himself published his *Fundamenta physices,* setting forth his own particular brand of Cartesian physics – but the cause had for the moment been lost, and the peripatetic tradition continued to dominate philosophic instruction at Utrecht through the middle of the century.[11] In the late 1640's, the center of the growing commotion over Cartesianism now settled at the University of Leiden, where the confrontation between the old and new philosophies would remain a source of acrid bitterness for several critical decades. A whole generation would be educated in a milieu permeated by conflict over philosophy and its theological ramifications. During this period, no discipline at Leiden would succumb so rapidly or so thoroughly to the new philosophy as the discipline of physics.

[9] See Serrurier, *Descartes,* and Elizabeth S. Haldane, *Descartes: His Life and Times* (London: John Murray, 1905), *passim.*

[10] Cited in G. J. Loncq, *Historische Schets der Utrechtsche Hoogeschool tot hare Verheffing in 1815* (Utrecht: J. L. Beijers en J. van Boekhoven, 1886), p. 50.

[11] Dibon, *La Philosophie néerlandaise,* Vol. I, pp. 218-9. Sassen, *Geschiedenis van de Wijsbegeerte in Nederland,* pp. 134-6.

In the intervening years between the death of Burgersdijck and the beginning of the troubles over Cartesianism at Leiden, no one of Burgersdijck's stature had succeeded him on the chair for natural philosophy. His immediate successor had been one Johannes Bodecher Benning, who had requested and received a temporary leave of absence after only three years in his new post.[12] Departing for Brazil in the service of the West India Company, he never returned to Leiden and died, apparently mad, in 1642. Better known as a poet – his satires had provoked student protests[13] – than a philosopher,[14] he left no writings relating to physics and appears to have had little if any impact on the orientation of natural philosophy within the university.

Upon Bodecher Benning's departure, the instruction in physics was entrusted to the professor of logic since Burgersdijck's death, François du Ban, a former Jesuit from France.[15] Du Ban, who also left no writings, remains problematic. Before coming to the United Provinces, he had taught at a number of schools in France, including La Flèche, where the youthful Descartes may have sat among his students. During du Ban's subsequent professorship at Leiden, the younger Adriaan Heereboord, later the most prominent defender of Cartesianism at the university, first joined the faculty, and it has been argued that it was in fact du Ban who first introduced Cartesianism at Leiden and converted his junior colleague.[16] There is nothing to indicate, however, that du Ban was a sympathizer of the new philosophy. The few theses on physics in surviving student disputations defended under du Ban are traditional; where they do deviate from the Aristotelian mold, it is not in the direction of Cartesianism.[17] Nor does du Ban appear to have dissented when the faculty, faced with the growing challenge of Cartesianism, reaffirmed the university's commitment to the traditional philosophy.

---

[12] On Bodecher Benning, see the *Nieuw Nederlandsch Biografisch Woordenboek*, Vol. IV, 178-9.
[13] *Bronnen*, Vol. II, pp. 154 and 160.
[14] Dibon, *La Philosophie néerlandaise*, Vol. I, p. 107.
[15] *Nieuw Nederlandsch Biografisch Woordenboek*, Vol. III, 58-9.
[16] Gustave Cohen, *Écrivains Français en Hollande dans la première moitié du XVIIe siècle* (La Haye: Martinus Nijhoff; Paris: Édouard Champion; 1921), pp. 336-7 and 653.
[17] See Hadrianus Pauw, *Disputatio philosophica, continens conclusiones aliquot, ex universa philosophia depromptas*, sub praesidio D. Francisci du Ban (Lug. Bat.: Ex officina Joannis Maire, 1637), and Philippus le Keux, *Disputatio philosophica, de fine*, sub praesidio D. Francisci du Ban (Lugduni Batavorum: Ex officina Bonaventurae et Abrahami Elsevir, 1640). Likewise, there occur no Cartesian theses in the *Disputatio philosophica* of J. Klenck, a student of du Ban's who graduated with this disputation in 1642 (Sassen, *Geschiedenis van de Wijsbegeerte in Nederland*, p. 148).

In 1641, when the troubles at Utrecht were mounting, the academic senate at Leiden requested that the curators authorize a commission of professors, including both du Ban and Heereboord, to draft an outline for the regulation of philosophic instruction.[18] The result of this commission, an *Ordo secundum quem deinceps in Academia Leidensi Philosophia docebitur,* was an unqualified confirmation of Aristotle in all branches of philosophy.[19] The text of Aristotle himself was to be read in the lectures and explained with the aid of classical interpreters. With regard to physics, the first four books of the *Physics* were to be read, the third book of *De caelo,* the two books composing the *De generatione et corruptione,* books one and four of the *Meteorologica,* and the second and third books of *De anima.* Of the members of the commission, only Heereboord delayed in appending his signature, and his reservations appear to have been less an expression of hostility towards Aristotle than a concern for pedagogical efficiency, hindered in the field of logic, he felt, by prescribing a verbatim (*verbotenus*) reading of the Aristotelian texts.[20] A former student and admirer of Burgersdijck, Heereboord appears to have been defending the importance of "method" in the teaching of philosophy, method that would have been undermined by a primary reliance on the reading of Aristotle's own works. Also counter to the inclination and practice of Burgersdijck and other prominent scholastic pedagogues was the exclusion by the *Ordo* of all but classical commentators.[21] Burgersdijck had made a point of the fact that, instead of the ancient commentators, he cited rather the more recent authors of the sixteenth and early seventeenth century, prominent among them being the representatives of the scholastic revival in Spain and Portugal, whose influence, it has been suggested, was itself a target of the Calvinist theologians.[22]

Heereboord's initial recalcitrance appears to have sprung primarily from his attachment to the pedagogical tradition exemplified by Burgersdijck, and he later signed a revised version of the *Ordo* which allowed for greater flexibility in teaching without, however, mitigating its exclusive Aristotelianism.[23] As for du Ban, there is no evidence that he ever had any compunctions about adhering to the *Ordo.* From Heereboord's

[18] *Bronnen,* Vol. II, p. 259.
[19] *Ibid.,* pp. 331*-3*.
[20] Dibon, *La Philosophie néerlandaise,* Vol. I, pp. 110-3. See Adriaan Heereboord, "Ad curatores epistola," in *Meletemata philosophica, maximam partem, metaphysica* (Lugduni Batavorum: Ex officinâ Francisci Moyardi, 1654), p. 9.
[21] See Reif, "Natural Philosophy," p. 26.
[22] *Idea philosophiae naturalis,* "Philosophiae studiosis." Dibon, *La Philosophie néerlandaise,* Vol. I, p. 113.
[23] Dibon, *loc. cit.*

testimony, we may gather that du Ban displayed a tolerance in philosophical matters that won the younger man's esteem – "We lived most closely joined in spirit," wrote Heereboord, "even though most often divided in our opinions in philosophy" [24] – but when du Ban died in 1643, Leiden's confrontation with Cartesianism was still to come.

The very project of the *Ordo* may already reflect a sense of anxious foreboding occasioned by the events at Utrecht, for no such formal prescription in philosophy had earlier been felt necessary in the statutes. Two years later, at du Ban's death, the uneasiness may be detected as well in the hesitation of the curators to appoint a new professor of physics. Having considered Heereboord himself as a possible successor to du Ban, they decided for the moment to grant only a provisional authorization as lector in physics to a Prussian doctor of medicine, Albert Kyper, and this on the explicit condition now, as had been recommended by the academic senate, that he remain within the limits of the traditional Aristotelian philosophy.[25]

Kyper, who had been teaching physics unofficially for several years,[26] was no intellectual radical. He adhered to the broad outlines of scholastic physics and in his own *Institutiones physicae,* published in two volumes in 1645, recommended as an aid to the reading of this text the *Collegium physicum* of Burgersdijck (as well as the *Physiologia Peripatetica* of Magirus and the *Philosophia naturalis* of Sennertus).[27] Though subscribing to Harvey's theory of the circulation of the blood, his medical views were also relatively conservative, and in later years, when a professor on the faculty of medicine and rector of the university, he evidenced his lack of sympathy for the Cartesian philosophy in particular by pointing out its injuriousness to the faculty of medicine.[28]

Nonetheless, Kyper's physics evidenced an independence and imaginativeness that doubtless contributed to his popularity among the students

---

[24] "Ad curatores epistola," p. 9.

[25] *Bronnen,* Vol. II, pp. 276 and 278. For more on Kyper's life, see G. C. B. Suringar, "Het geneeskundig Onderwijs van Albert Kijper en Johannes Antonides van der Linden. De ontleedkundige School van Johannes van Horne," in Suringar's *Bijdragen tot de Geschiedenis van het Geneeskundig Onderwijs aan de Leidsche Hoogeschool, van de Stichting der Universiteit in 1575 tot aan den dood van Boerhaave, 1738* (twaalf opstellen, overgedrukt uit het *Nederlandsch Tijdschrift voor Geneeskunde,* jaargang 1860-1866). See also the *Nieuw Nederlandsch Biografisch Woordenboek,* Vol. II, 732.

[26] See *Bronnen,* Vol. II, p. 250, and the "Epistola dedicatoria" and "Praefatio ad lectores" in Kyper's *Institutiones physicae* (Apud Franciscum Moiardum et Adrianum Wyngaerden, 1645).

[27] *Ibid.*

[28] Suringar, "Het geneeskundig Onderwijs van Albert Kijper," *passim. Bronnen,* Vol. III, p. 107.

of philosophy who, only months before he was granted his lectorship in physics, had petitioned the curators that he be allowed to preside over public disputations.[29] "Everywhere," he said of his rambling and idiosyncratic *Institutiones,* "I have followed my own wit, for I always detest servitude." [30] As a consequence, no doubt, he felt obliged to defend himself against "clandestine whispers and false accusations" charging him with cultivating new opinions in philosophy dangerous to the foundations of theology and the peace of the university.[31]

Within the next year of Kyper's authorization, indeed, the curators had found a new professor, a staunch Aristotelian, to assume the empty chair in physics, and two years later Kyper himself temporarily departed from Leiden to become professor of natural philosophy and medicine at the Illustre School at Breda and personal physician to the Prince of Orange. His moment of responsibility for physics at Leiden was brief, and his impact on the development of natural philosophy there cannot be considered of any great significance. Nevertheless, his physics represents a version of the "traditional" philosophy apparently popular among the students, a version also more strongly marked now by the corrosive influence of the recent scientific developments.

Perhaps most provocative and dramatic in Kyper's physics was the accelerated disintegration of the scholastic cosmos. Like Burgersdijck, Kyper rejected the crystalline spheres and declared it more probable that the heavens were one continuous fluid body.[32] Unlike Burgersdijck, however, he now identified this celestial fluid with the air of the sublunar world and abandoned, thereby, the dichotomy between the celestial and the terrestrial realms. "There is no evidence," he declared, "by which it is possible to demonstrate that the heavens are necessarily of a different nature than the elements." [33] As for the ostensible incorruptibility of the heavens, he observed that the senses were unable to detect any generation or corruption in the air itself just as, on the other hand, they were unable to detect an immunity to these processes in the heavens.[34] Nor were the heavens devoid of gravity and levity, the rectilinear natural motions that had previously distinguished the sublunar elements:

Indeed, imagine with me that the heavens were raised up so that they were no longer united with the other bodies below. Would nothing move downwards as rapidly as possible, that is, in a straight line, towards union again? Or if some part

[29] *Bronnen,* Vol. II, p. 275.
[30] *Institutiones physicae,* "Praefatio ad lectores."
[31] *Ibid.,* "Epistola dedicatoria."
[32] *Ibid.,* Vol. II, pp. 9-10.
[33] *Ibid.,* Vol. I, p. 391.
[34] *Ibid.,* Vol. I, pp. 391-2; Vol. II, pp. 6-7.

of the heavens were submerged below the water, would nothing strive to rise directly upward? [35]

Having also asserted the presence in the heavens of the primary qualities – heat, cold, wetness and dryness – traditionally coupled with the sublunar elements, he concluded that "the heavens and the air are in essence one and the same body . . . ." [36] This he confirmed with a closing argument drawn from the Scripture. With Biblical citations for support, he declared that the expanse created on the second day of the Creation had comprised both the heavens and the air: "Therefore, the air is the heavens, or, by transposition, the heavens are the air." [37] Indeed, while illustrating the inroads being made into the fabric of scholastic physics, Kyper also testified to the continuing authority of the Bible as a source, in many circles, of irrefutable natural as well as spiritual knowledge.

The stars now also partook of the substance of the sublunar world. Again combining scriptural and philosophical arguments, Kyper explicitly denied the notion, to which Burgersdijck had adhered, that the stars were but denser portions of the heavens.[38] Genesis seemed to say that God had placed the stars within the heavens, not that he had made them from the heavens; the heavens and the air, again, were one continuous body, "but no one would assert that the stars are denser portions of the air." [39] The stars, after all, were mixtures of the elements, and though fire might frequently be predominant, the other elements were doubtless present, as spots on the sun and moon and the differences in stellar light seemed to indicate. In Kyper's thought, the cosmos had come much closer to a homogeneous universe in which physical substance was everywhere the same and everywhere subject to the same physical processes.

Braced by the authority of the Bible, however, Kyper shared none of Burgersdijck's uncertainty over Copernicanism, for the Holy Scripture, Kyper declared, explicitly taught that the stars moved and the earth stood still.[40] He added to his scriptural protest a no less common objection based on what he believed would be the consequences of the movement of the earth: cannon balls would not fly equally towards the east and west, the flight of birds would appear similarly distorted, houses would collapse and so forth. Closely related to the problem of projectiles, this criticism of the Copernican theory in terms of terrestrial effects

[35] *Ibid.*, Vol. I, p. 391.
[36] *Ibid.*, Vol. I, p. 392.
[37] *Ibid.*
[38] *Ibid.*, Vol. II, p. 19 ff.
[39] *Ibid.*, p. 19.
[40] *Ibid.*, p. 104; see also p. 57.

would also be resolved only by a new understanding of motion. As for the argument for a moving earth that Burgersdijck had found so compelling – that the stars and planets would have to be moving at incredible speeds if the earth were at rest – Kyper observed that the speed of the earth itself if it, on the other hand, were moving would also be beyond the grasp of reason, some 13,760 miles an hour he reckoned.[41] Kyper himself adhered to the system devised by Tycho Brahe, in which the earth remained at rest and at the center of a universe which was otherwise mathematically identical to the system of Copernicus. Only the moon and sun orbited directly about the earth, the other planets orbiting about the moving sun.[42] In the course of the century, this Tychonic system replaced the Ptolemaic as the major refuge of those who still clung to an unmoving earth at the center of the universe, but with its great secondary orbits about the sun, Tycho's system also shattered the simple symmetry of the Aristotelian cosmos.

With the progressive disintegration of the physical innards of the cosmos, the imagination was loosed to cast about for new alternatives to explain the persisting cosmic order. As was so often the case, Kyper's own conjectures combined traditional ideas and fallacious interpretations of new observations with provocative suggestions of the future.

> It is considered doubtful that, besides the motion about the earth, there is also motion proper to the stars which can be attributed to gravity or levity. In truth, I consider it not improbable. We have suggested above that there is some particular gravity and levity which might be referred to with respect not to the center of the earth but to the center of each individual body. Some examples of this seem to be observed in the stars.[43]

Among these examples were the solar and lunar spots, which Kyper spoke of as tending towards (*vergunt ad*) their respective celestial bodies. He mentioned the "circle of vapors" which appeared to be hovering around the moon during eclipses and the similar vaporous envelopment which the varying brightness of Jupiter's satellites (here he was echoing the suggestion of Galileo himself in the *Sidereus Nuncius*) revealed about that planet as well. Earlier, he had asserted that the complex but enduring motions of the stars – the planets, that is – could be derived from sympathies and antipathies between the stars themselves and lower bodies.[44] Though surely too crude to be considered an anticipation of Newtonian gravitation, Kyper's speculations bespoke nonetheless the crumbling

---

[41] *Ibid.*, p. 107.
[42] *Ibid.*, p. 37.
[43] *Ibid.*, p. 110.
[44] *Ibid.*, p. 29.

coherence of the cosmos and the need for radical new expedients to account for the continuing order of the heavens.

It was the practice of medicine, not the teaching of philosophy, to which Kyper ultimately aspired,[45] and his natural philosophy does not reflect the pervasive concern for systematization that characterized a philosopher-pedagogue like Burgersdijck. The *Institutiones physicae* is the product of a mind too readily intrigued by a wide diversity of questions and issues and relatively indifferent to the need for a unifying philosophic orientation; "the unity of physics," he himself declared, "is not essential or specific, but is the unity of an ordered aggregation." [46] Consequently, his physics bespoke the great variety of problems that might still be considered pertinent to a science of nature. In addition to seeking a new explanation for the structure of the cosmos, he also asked whether man or the stars were more perfect and if Christians could, with good conscience, retain the old names of the stars.[47] He described simple experiments he may well have performed, while elsewhere he puzzled over such questions as whether darkness was to be considered as having real being.[48]

Kyper's diffuse but imaginative approach, we may assume, was stimulating to many students, as was his critical dissatisfaction with the philosophical systems of others (partly, he confessed, because he could not always understand their thinking) and his determination, hence, to develop a system of his own.[49] In those years, however, when the emerging challenge of Cartesianism had intensified the distrust of any innovation, Kyper's indulgence of his "own wit," no matter how far he strayed from what we might now call the spirit of modernity, would inspire little confidence among those who had determined to prevent the University of Leiden from becoming a battleground of academic strife.

Having considered several candidates with regard particularly to their adherence to the Reformed religion and their avoidance of "novelties," the curators finally offered the professorial title in physics to the Scot, Adam Stuart, in late 1644.[50] Then resident in London, Stuart had taught at the Huguenot Academy at Sedan and had years earlier been considered as a possible successor to Burgersdijck.[51] A conservative Aristotelian, he was recommended by the faculty of theology, who also suggested that

---

[45] *Ibid.*, "Praefatio ad lectores."
[46] *Ibid.*, Vol. I, p. 13.
[47] *Ibid.*, Vol. II, pp. 46 and 42.
[48] *Ibid.*, Vol. I, pp. 314 ff., 394 and 463.
[49] *Ibid.*, "Praefatio ad lectores."
[50] *Bronnen*, Vol. II, pp. 276 and 287-8.
[51] *Ibid.*, pp. 287 and 195.

he be entrusted, as he was, with the first public instruction in metaphysics offered by the university.[52] The new title of *professor primarius* of philosophy was further created for him so that he should have rank over the younger Heereboord.[53] Vigorously supported by the *professor primarius* of theology, Jacobus Triglandius, and the regent of the state college for theology students at the university, Jacobus Revius,[54] Stuart took up arms against the advance of Cartesianism at Leiden and was soon deeply embroiled with the enthusiastic defenders of the new philosophy, chief among them being his junior colleague, Heereboord.

A former student of theology at Leiden, Heereboord had been with the faculty of the university since assuming the professorship of logic in 1641.[55] Subsequently to acquire the chair in ethics as well, he was also named subregent of the theological college, placing him directly under his future adversary, the regent Revius.[56] That he was a potential source of discord soon became apparent. When Stuart made his appearance in 1644, Heereboord's students were already debating theses of Cartesian inspiration, and in the following year, in a public address, Heereboord lauded Descartes himself as a "morning-star" who had displayed the key to philosophy and revealed the way to unshakable truth.[57]

In 1646, Triglandius protested to the academic senate about a thesis defended under Jacobus Golius, professor of mathematics and an acquaintance of Descartes, that maintained that the quest for certain knowledge was to begin with doubt.[58] Such theses smelled of heresy, declared the theologian, and the senate, in accord with the wishes of the curators, promptly responded by reaffirming that only the philosophy of Aristotle was to be taught at the university.[59] In January of the following year, however, Heereboord delivered yet another address, now on the "freedom of philosophizing," in which he described Descartes as "unequalled in the cause of truth emerging from darkness and servitude" and as pointing the way, if we would but heed him, to a philosophy free

---

[52] Sassen, *Geschiedenis van de Wijsbegeerte in Nederland*, p. 149. Schotel, *De Academie te Leiden*, p. 159. *Bronnen*, Vol. II, pp. 295-6. Metaphysics, nonetheless, had unofficially been taught at Leiden on and off since at least the beginning of the century (Sassen, *op. cit.*, p. 124).
[53] *Bronnen*, Vol. II, p. 287.
[54] Siegenbeek, *Geschiedenis der Leidsche Hoogeschool*, Vol. I, p. 161.
[55] *Bronnen*, Vol. II, pp. 246 and 251-2. See Sassen, "Adriaan Heereboord (1614-1661), De Opkomst van het Cartesianisme te Leiden," *Algemeen Nederlands Tijdschrift voor Wijsbegeerte en Psychologie*, Vol. XXXVI (1942-3), pp. 12-22.
[56] *Bronnen*, Vol. II, pp. 296 and 267.
[57] Heereboord, "Ad curatores epistola," pp. 9-11.
[58] *Ibid.*, pp. 12-3. Serrurier, *Descartes*, pp. 47 and 181 ff.
[59] Serrurier, *loc. cit.*

of the prejudices that now obstructed it.[60] Revius and Triglandius soon responded with disputations in which Descartes was accused of blasphemy and Pelagianism, an attack which occasioned not only further exchanges within the faculty but a lengthy letter of protest to the curators from Descartes himself.[61]

The curators, anxious to stifle the bitter controversy that was now growing within and around the university, reprimanded both Revius and Heereboord (who had Stuart called in as well) and prohibited the further mention of Descartes' name or ideas in the lessons, disputations or other public activities of the professors.[62] This prohibition, first proclaimed in 1647, was to remain the consistent policy of the curators in their effort to calm the turmoil that continued to plague the university throughout the following years. It was a policy, however, that won few adherents among the professors of either faction. Though Descartes again protested to the curators that their action prevented his defending himself, Heereboord continued to champion his cause with a bristly aggressiveness that became notorious.[63] Stuart, in his turn, declared that he could keep silent about the "atheistic" writings of Descartes only with the "greatest offense to God, grave scandal to the Church, and the eternal damnation of my soul." [64] Revius, likewise, continued a vehement campaign against Descartes and his supporters despite repeated admonishments from the curators.[65]

The continuing and increasingly acrid feud between the professors was seconded by the turbulent partisanship that now divided the student body.[66] Already in 1648 Triglandius was complaining that the young students were all abandoning the disputations in theology and flocking to those in philosophy whenever the latter touched on "novelties," and Heereboord was later to pride himself on having made the disputations in philosophy more exciting than any others.[67] So exciting were they that in February of 1648 a disputation under Stuart was interrupted by a fist fight, and a few months later Stuart complained to the curators of the continuing efforts of Heereboord and a young doctor of medicine at Leiden, Joannes de Raey, to incite the "Cartesian sect" to disrupt his

[60] "Ad curatores epistola," p. 13.
[61] René Descartes, *Oeuvres*, ed. Charles Adam and Paul Tannery (Paris: Léopold Cerf, 1897-1910), Vol. V, pp. 1-15.
[62] *Bronnen*, Vol. III, pp. 4-6.
[63] Descartes, *Oeuvres*, Vol. V, pp. 35-9. Thijssen-Schoute, "Le Cartésianisme aux Pays-Bas," p. 212.
[64] *Bronnen*, Vol. III, pp. 13* and 17*.
[65] Sassen, *Geschiedenis van de Wijsbegeerte in Nederland*, pp. 160-1.
[66] Thijssen-Schoute, "Le Cartésianisme aux Pays-Bas," p. 219.
[67] *Bronnen*, Vol. III, p. 16. Heereboord, "Ad curatores epistola," p. 9.

lecture hall by any means.⁶⁸ The student harassment of Stuart, whatever its source, became such, indeed, that he abstained from presiding over public disputations for the next two years, until instructed by the curators in 1651 to carry out all his expected duties.⁶⁹ In 1653, consequently, he was again complaining of outbursts in his classes and insults from the students (and from colleagues in the academic senate as well).⁷⁰ In the following year, death finally relieved Stuart of his trials, but student disruptiveness continued, largely directed now at Henricus Bornius, a former student of Heereboord's who had nonetheless acquired a reputation as an anti-Cartesian since joining the faculty of philosophy in 1652.⁷¹ In 1658, however, it was the rector of the university himself, Antonius Thysius, professor of eloquence, who complained of student harassment attributed to the influence of Heereboord.⁷² The baiting of professors at Leiden had become so notorious, indeed, that in the following year, 1659, the States of Holland and West Friesland were moved to issue an edict expressly forbidding stamping and banging in lectures, orations or disputations.⁷³

Three years earlier, the States had already intervened on a somewhat higher plane to try to quell the turmoil at the university. Spurred on by complaints from ecclesiastical bodies within the province, the States had consulted with the curators, the faculty of theology, the rector, and other members of the university and had issued, in 1656, a decree against mixing theology with philosophy and abusing the freedom of philosophizing to the prejudice of the Scripture.⁷⁴ It was specifically pointed out by the States that their resolution was not to be interpreted as weakening in any way the prohibitions of the university curators against references to Descartes or his philosophy. The curators themselves were instructed to do all that they could to see that the professors did not abuse the freedom of philosophizing to the detriment, as well, of the mutual affection and friendship that should have prevailed among the professors as members of one and the same body.

It was customary that theses to be defended were printed and disseminated as a preliminary announcement of an impending disputation, and less than two months after the promulgation of the decree of the

⁶⁸ *Bronnen,* Vol. III, pp. 10, 16 and 13*.
⁶⁹ *Ibid.,* pp. 55-6.
⁷⁰ *Ibid.,* p. 73.
⁷¹ Sassen, "Adriaan Heereboord," p. 20; *Geschiedenis van de Wijsbegeerte in Nederland,* pp. 149-50. Thijssen-Schoute, "Le Cartésianisme aux Pays-Bas," p. 235.
⁷² *Bronnen,* Vol. III, p. 141.
⁷³ de Vrankrijker, *Vier Eeuwen,* p. 65.
⁷⁴ *Bronnen,* Vol. III, pp. 55*-8*.

States, to which the professors of philosophy had been obliged to take an oath, the great statesman of the republic during these years, Johan de Witt, was personally protesting to the rector of the university about a disputation soon to be defended under Heereboord.[75] De Witt found the invidiousness of certain theses maintaining that Aristotelianism was antagonistic and injurious to the deity inexcusable, and he called for the cooperation of more tactful sympathizers of the new philosophy at Leiden in helping to ensure that "the imprudence of one man, who presents himself as the defender of the just liberty of philosophizing, might not be used by others to abridge the freedom of all reasonable philosophers." [76]

Heereboord's prominence in the struggle over Cartesianism, however, was soon to fade.[77] Age had clearly not brought docility, for in 1657 he antagonized the faculties of theology and philosophy at Utrecht with the announcement of a disputation in which it was suggested that they were guilty of heterodoxy, and in 1658 both Thysius, as we have seen, and Bornius were complaining of the injuries done them by their colleague.[78] Nonetheless, the cause of Cartesianism at Leiden was now being advanced by more genuine Cartesians, and with regard to physics in particular, Heereboord had fallen well behind the progress of the new philosophy within the university.

Since the beginning of the hostilities at Leiden, passions there as elsewhere had been most inflamed by the involvement of theology. Apart from the participation of the theologians themselves, the first clashes between Heereboord and Stuart had turned about theological as well as philosophical issues,[79] and Stuart felt that to abandon his resistance to Cartesianism was to endanger his spiritual well-being. As early as 1648, the faculty of theology as a whole had protested the incursions of both Stuart and Heereboord into matters pertinent to theology,[80] and the mingling of philosophy and theology had remained the primary concern of the States of Holland and West Friesland in their decree of 1656. Discord over metaphysics also flourished, and the curators, arguing that the subject engendered excessive partisanship, revoked Stuart's authorization to teach the subject in 1648.[81] Stuart regained this authorization

---

[75] *Ibid.*, p. 63*. Thijssen-Schoute, "Le Cartésianisme aux Pays-Bas," pp. 210 and 212.
[76] *Bronnen*, Vol. III, pp. 64*-5*.
[77] Sassen, "Adriaan Heereboord," p. 20.
[78] *Bronnen*, Vol. III, pp. 65*-6* and 141.
[79] Heereboord, "Ad curatores epistola," pp. 11-2.
[80] *Bronnen*, Vol. III, pp. 16*-7*.
[81] Sassen, "Adriaan Heereboord," p. 17.

in 1653, subject now, however, to "certain precautions" that the curators do not appear to have elaborated in writing.[82] Heereboord himself had also sought permission to teach metaphysics and even without such authorization felt no greater compunctions about trespassing in metaphysics than in the province of the theologians.[83]

Both philosophers had turned to physics as well. Stuart remained responsible for that discipline until 1653, and the most independent of Heereboord's writings, though he never held the chair in physics, was a posthumous *Philosophia naturalis* published in 1663, two years after his death.[84] In 1654, Heereboord had also published a *Philosophia, naturalis, moralis, rationalis* in which the philosophy and the organization of his physics remained very much in the scholastic tradition of Burgersdijck.[85] Three years before his death, however, he was pressed by students to elaborate this earlier work on physics and to incorporate both the opinions of more recent philosophers and his own thoughts where he dissented from the others.[86] The result was the posthumous text, his earlier scholastic text now augmented with new commentaries reflecting his own views and those chiefly of Descartes, Regius, and Claude Bérigard, whose *Circulis Pisanus* (1643, 1661) had opposed to Aristotelian physics the philosophy of the Ionian pre-socratics.

It was an awkward format; the new commentaries were interspersed throughout the scholastic theses and their original scholastic commentaries.[87] The reliance on the older framework as well as the disparate character of the new commentaries reflect a failure on Heereboord's part to have worked out a consistent and systematic physics of his own. The "progressive" character of his physics remained largely a diffuse and eclectic criticism of the older system, for which, however, since he could not accept Cartesian physics as a whole, he had no viable alternative.

---

[82] *Bronnen*, Vol. III, p. 76.

[83] Sassen, "Adriaan Heereboord," p. 18. Heereboord, "Ad curatores epistola," p. 10. A great part of Heereboord's writings, indeed, pertained to metaphysics.

[84] *Philosophia naturalis, cum commentariis peripateticis antehac edita: nunc vero hac posthumâ editione mediam partem aucta, et novis commentariis, partim è Nob. D. Cartesio, Cl. Berigardo, H. Regio, aliisque praestantioribus philosophis, petitis, partim ex propria opinione dictatis, explicata* (Lugduni Batavorum: Ex officinâ Cornelii Driehuysen, 1663).

[85] Lugduni Batavorum: Ex officinâ Francisci Moyardi, 1654.

[86] *Philosophia naturalis cum novis commentariis explicata*, "Typographus benevolo lectori S."

[87] The work appeared in a somewhat different form in the volume of his *Meletemata philosophica* published in 1665 (Neomagi, Ex officinâ Andreae ab Hoogenhuysen), in which the newer commentaries were included together as a separate section following the scholastic text.

Though rejecting certain fundamental concepts of scholastic physics, he was unable to do without the systematization it provided.

Heereboord's effort in natural philosophy is of interest not only because of the problems it raises concerning the nature of his disruptive influence at Leiden, but also because of the insight it provides into the difficulty of the philosophic journey on which the century had embarked. His physics exemplified the increasing dissatisfaction with basic scholastic formulations and the welcome which was accorded to the alternatives now offered by the new philosophy. On the other hand, it illustrated no less the formidable obstacles which the advance of seventeenth-century science placed before even the fervent defenders of innovation and philosophic reform.

Perhaps the clearest and most unambiguous note struck in the new commentaries of the *Philosophia naturalis* was the denial of the Aristotelian first principles. Privation was rejected as a logical absurdity [88] – "non-being is not a principle of being," Kyper had also asserted [89] – and peripatetic first matter, existing only in man's reasoning, was declared incapable of being a constitutive component of natural bodies: "bodies are real things, which cannot be composed from a thing of reason as if it were some essential part." [90] A first matter which had neither quality nor definite quantity could not be anything at all.[91] "Therefore, if it is neither body nor spirit, there can be no concept of first matter, that is, it can not be conceived. Therefore, indeed, I allow that it has never existed, much less as a principle of natural bodies." [92] Heereboord testified here to the longing for a reality in the principles of natural bodies more suggestive of physical existence itself, a longing which his master Burgersdijck and other late scholastics had patently failed to satisfy. If matter was not conceived in terms of such attributes as figure, motion, magnitude and position, continued Heereboord, there would be no concept of matter at all, or a concept, at least, that would remain extremely confused.[93]

Heereboord referred in the latter part of his comments on first matter to his three modern philosophers, and the attributes he required of a conceivable matter were the attributes of matter in the mechanical philosophy. The emphasis on conceivability as a criterion of reality was perhaps itself evidence of the Cartesian influence in Heereboord's

[88] *Philosophia naturalis cum novis commentariis explicata*, pp. 8-9.
[89] *Institutiones physicae*, Vol. I, p. 54.
[90] *Philosophia naturalis cum novis commentariis explicata*, p. 11.
[91] *Ibid.*, pp. 15-21.
[92] *Ibid.*, p. 18.
[93] *Ibid.*, p. 19.

thought, for in his determination to lay a foundation of certain knowledge, Descartes had stressed that only concepts that were clear and distinct were to be accepted as true. When applied to the realm of corporeal existence, this criterion tended to become a test of crude imaginableness, an appeal, despite the Cartesian denigration of the reliability of the senses, to a penetrating vision and sense of touch, as it were, within the mind. The immediacy, simplicity, and seeming completeness of such conceptualization was not merely a welcome relief but an inspiring reassurance after the abstruse logical complexities of the scholastics. It was the most compelling attribute of the mechanical philosophy.

Heereboord declined, however, to accept the fundamental Cartesian identification of matter with pure extension. Although he did suggest at one point that "substance extended in length, breadth and depth" might be a perfect definition of first matter,[94] he elsewhere remarked in passing that Cartesian body was not to be understood as complete body, physical body, but as that part of body that was called mathematical body.[95] By continuing to adhere to the distinction between the physical and mathematical, Heereboord rejected the basic premise wherein Cartesian physics would prove to have overreached the new science itself. Where he paused to consider the question at greater length, however, it became apparent that it was the scholastic tradition that had most shaped his thinking. He declared that, like Burgersdijck, he adhered to the position of the Scotists, who considered extension a property of matter distinguished from matter by reason.[96] He found the Cartesian identification of matter and extension difficult to accept for two reasons. First, it was precisely its extension that distinguished matter from spirit, and a thing itself should be different, Heereboord maintained, from that which distinguishes it from something else. Further, the Cartesian definition, "matter is extension, body is quantity," was better said: matter is extended, body is quantitative (*corpus est quantum*). "Who has ever said, 'man is rationality,' and not rather, 'man is rational'?" [97] And extension stood in the same relation to matter, he assured his reader, as rationality to man.

In addition to his repudiation of privation and first matter, Heereboord likewise had little patience now with the concept of form, particularly, except when it concerned the soul of man, the substantial form as expounded by late scholastics like Burgersdijck. Implicitly assuming, again,

---

[94] *Ibid.*, p. 15.
[95] *Ibid.*, p. 81.
[96] *Ibid.*, pp. 21-3 and 46-8.
[97] *Ibid.*, p. 47.

that the principles of physics should conform, at least to some minimal degree, to what was expected of physical existence itself, Heereboord derided what he represented as the teachings on the generation of form. "How is form brought forth from the potentiality of matter?," he challenged; "How does it lie so long concealed therein before it rises and comes forth?" [98] The derivation of the forms of material things from potentiality was nothing other than the creation of something from nothing, he asserted, "and thus vision cannot be said to be elicited from the potentiality of a blind eye deprived of this vision unless we also say that this vision is elicited from its own nothingness (*ex nihilo sui*)." [99]

Heereboord's major complaint against the scholastic explication of form was that form had become something substantial in itself and distinct from matter, a criticism, he pointed out, in which Aristotle himself might well have concurred.[100] Consequent to this first error, Heereboord perceived a second, that matter was conceived as completely passive and inactive (*iners*), all the activities of natural bodies being derived from the substantial form alone. Although matter itself was also totally inert in the mechanical philosophy, it was the influence of the mechanical philosophy, the conviction that the phenomena of nature resulted ultimately from matter in motion, that lay behind Heereboord's second complaint.

Though rejecting the Cartesian definition of matter as pure extension, Heereboord eagerly embraced Descartes' corpuscular imagery. He was clearly enthusiastic about the depiction of a realm of "parts" of extended matter so small as to be unfelt and unseen, yet which, through the sum of their motions, produced the sensible appearance of the world.[101] Rich with the delights of imaginative fantasy, this imagery was to prove a seductive feature of Cartesian physics and the mechanical philosophy in general. It played a decisive role, as well, in sealing the fate of the peripatetic doctrine of form.

Heereboord had asserted in an earlier scholastic commentary that the existence of substantial forms was apparent from the different combinations of accidents to be found in different things.[102] In a new commentary, however, he answered that these differences in accidents had another cause than the substantial form, and that was "motion, either greater or less, with either more, that is, or less rapidity, in the parts of

[98] *Ibid.*, pp. 25-6.
[99] *Ibid.*, p. 29.
[100] *Ibid.*, pp. 25-6.
[101] *Ibid.*, see pp. 22-3.
[102] *Ibid.*, p. 26.

matter...."[103] If this were so, if all the activities and properties of natural bodies could be explained by the motion of parts of matter, why, he asked, need we accept form as a substance distinct from matter?[104] The representation of the substantial form as itself an independent substance continued to provoke Heereboord before all else; ironically, the efforts of the late scholastics to endow the principle of form with greater reality had left it more exposed and less tolerable to their heir. The corrosiveness of the corpuscular imagery, however, undermined more than just the substance of the substantial form. The variety of accidents perceived in natural bodies no longer necessarily testified to form at all or to potentiality and actualization; "it is enough," wrote Heereboord, "that a natural body is composed of diverse but integrated (*integrantibus*) parts of matter differing in magnitude, figure, position, proportion, distance, motion, etc."[105]

These, indeed, were only accidents of matter, but all together, stressed Heereboord, these particular accidents provided natural bodies with their specific natures and distinguished them one from another;[106] it was these accidents, that is, that now made a natural body what it was, that determined its identity. Heereboord illustrated his point with a popular image in the iconography of the mechanical philosophy, the clock:

...in which the motion, position, figure and magnitude – all of which, indeed, occur in matter – are accidental to the iron but essential to the total clock. Without the disposition of parts as is found in a clock, without this determination of magnitude, without this motion, position, figure, etc., it is certainly iron and endures as iron, but it is never a clock or endures as a clock.[107]

The consciousness of the transient identities in nature was now being compromised by the idea of a universal and perpetual flux of accidents, and, consequently, the meaningfulness of the scholastic concepts of generation and corruption was threatened as well. Heereboord explained the position of his three "moderns": since they recognized no substantial form (apart from that of man), they recognized no substantial generation (except man's) and considered generation nothing but the motion and change of parts of matter.[108] No new substance was produced, "but only a new modification of this same substance, that is, of matter."

Despite a passing reservation, Heereboord's acceptance of this mecha-

[103] *Ibid.*, pp. 26-7.
[104] *Ibid.*, p. 28.
[105] *Ibid.*, p. 34.
[106] *Ibid.*, p. 36.
[107] *Ibid.*
[108] *Ibid.*, pp. 124-5.

nistic negation of the older concept is evident elsewhere,[109] but in his own new commentary on the original scholastic theses, he himself rejected generation and corruption in a manner quite foreign to the new philosophy.[110] Repudiating the thesis that in generation and corruption no perceptible subject remained throughout, or if it did, that it differed in species, Heereboord directed his argument towards the death of man, the corpse of which, the scholastics were said to have taught, was a different species than the living body. To oppose this line of reasoning, Heereboord appealed to the doctrine of the resurrection. In denying, as well, that generation and corruption consisted of the production and destruction of form, he cited the rational soul, man's substantial form, which was neither produced by the human father, he asserted, nor destroyed at death. Heereboord's arguments reveal how great the distance still was, despite his mechanistic sympathies, between his natural philosophy and the new science that was taking shape.

Heereboord also declined to resort to the mechanical imagery on other occasions, even when it offered a ready solution to certain venerable dilemmas within the scholastic tradition. When reconsidering what happened to the forms of the elements in a mixture – "a most agitated question among Aristotle and his Greek, Arab and Latin interpreters" – he did not stray from the traditional context.[111] He cited Avicenna, Averroës and Zabarella and spoke of the elemental forms no differently than had Burgersdijck – "that they do not perish is certain, for nothing passes into nothing naturally." The relevant views of Descartes, Regius and Bérigard were outlined in a special section that followed,[112] but they found no echo in Heereboord's own personal re-examination of the problem. The similar question of the "temperament" of a mixture was likewise handled without mention of the parts of matter in motion, which were left again to the vivid but noncommittal résumé of the three moderns at the end of the chapter.[113]

The insensible particles had again played a significant role, however, in the small progress Heereboord had made towards the new understanding of motion. Descartes and Regius, observed Heereboord, held all motion, including generation and corruption as well as accretion, diminution, and alteration, to be only local motion, and Heereboord agreed: "and surely, if we consider the matter truthfully, neither reason nor the senses dictate that anything besides local motion is to be admitted in any

[109] *Ibid.*, pp. 125 and 80-1.
[110] *Ibid.*, p. 115 ff.
[111] *Ibid.*, pp. 142-3.
[112] *Ibid.*, p. 150 ff.
[113] *Ibid.*, pp. 146 and 150 ff.

of these...." [114] Both accretion and generation (for which Heereboord here accepted the mechanistic rendering) were but the parts of matter coming together and solidifying in a composite body, while diminution and corruption were, correspondingly, the subsequent separation of those parts. Heereboord's explanation of alteration, however, became somewhat anomalous. Although he elsewhere put forward the more purely mechanistic understanding of heat as but the more rapid and agitated motion of the particles of matter,[115] he now declared that heating and cooling, though consisting only of local motion, were the addition and expulsion of degrees (*partes graduales sive gradus*) of heat. He likewise provided a perversely uninformative account, ostensibly Cartesian, of the qualities of wetness and dryness: wetness, he wrote, was the presence of sensible particles of water or liquid in the pores of bodies, and dryness was their absence. The recognition of all change as local motion was only the beginning of the Cartesian reconceptualization of motion, and Heereboord had not even got the beginning quite right. The more difficult conceptual somersaults that were yet to come Heereboord either could not or would not perform.

Descartes and Regius, Heereboord informed his readers, considered all motion to be both natural and violent at the same time, natural in that all motion conformed to laws posited in nature by God, violent in that the cause was always external to the body moved.[116] To the moderns, it was violent motion not only when a stone was thrown upwards into the air, with which the Aristotelians would concur, but when it was dropped from the hand as well, a motion which to the Aristotelians was purely natural, deriving from form, from the internal principle of heavy bodies. Heereboord himself, however, was unwilling or unable to part with the Aristotelian distinction between natural and violent motion. In his own new commentary to the old scholastic thesis, apart, that is, from his résumé of the moderns, he concentrated exclusively on distinguishing between the meaning of the word "violent" in physics and ethics.[117] It was the fundamental Aristotelian conception of natural motion that he could not bring himself to abandon.

Consequent to his argument that all motion was local motion, Heereboord had observed that all the remaining scholastic theses on motion, if they were not now completely undone, continued to be applicable to local motion itself:

[114] *Ibid.*, pp. 80-1.
[115] *Ibid.*, p. 27; see also p. 153.
[116] *Ibid.*, pp. 82-3.
[117] *Ibid.*, p. 76.

... what is said, surely, concerning the clear distinction there should be between the mover, the movable, the *terminus a quo,* and the *terminus ad quem* is true if understood with regard to local motion alone. For when a body is transferred from place to place, or from the vicinity of some places to the vicinity of others [an echo of the Cartesian statement of the relativity of motion], the *terminus a quo* is distinct certainly from the *terminus ad quem,* the place *from which* it is transported from the place *to which* it is transported, and the body which is itself transported is distinct, as well, from that which transports it.[118]

In another passage of the new commentaries, resorting to a traditional illustration of motion, he reaffirmed the close special relationship that existed between motion and the end towards which it moved.

Among these four, necessary for every motion, none is more intimate to motion than the *terminus ad quem,* for this is identified with motion, [but] not so the mover, movable, and *terminus a quo,* which are always distinguished from motion itself. But the *terminus ad quem,* such as heat, is not distinguished from the motion, heating, except through reason alone.[119]

Heereboord had not freed himself from the teleological perception of motion, wherein the very significance of motion derived from the goal it necessarily sought.

When he took up the motion of the heavens, Heereboord fell back on the peripatetic concept of natural motion in its purest sense, even though he turned it against Aristotle himself. In his earlier commentary, Heereboord had declared that the motions of the heavens were not to be attributed to "intelligences," as Aristotle, he said, had done, but to the form of the heavens, and Heereboord continued to maintain this point in his new commentary as well.[120]

Thesis 47 is opposed to Aristotle but not to the truth of the matter, and thus it can be proved: whatever is a natural body possesses its own principle of motion and not from without but within, for nature is the principle and cause of motion in a natural body; the heavens, however, are a natural body, *ergo &c.*

Heereboord had spoken of the Aristotelian definition of nature as "astonishing, not in its elegance or perspicuity, but in its difficulty and obscurity"; [121] nevertheless, it continued to shape Heereboord's understanding of motion.

In the process of reducing the substantial form to but a multitude of tiny particles of matter in motion, Heereboord had, in passing, suggested a doctrine of motion hardly compatible with that of natural, teleological motion. "If you ask from whence comes that motion and from what

[118] *Ibid.,* pp. 81-2.
[119] *Ibid.,* p. 66.
[120] *Ibid.,* p. 94.
[121] *Ibid.,* p. 40.

principle," he had written, "I answer that that motion was placed and implanted in matter and its parts from the first creation and that it never ceases, but is greater in one body or element and less in another...." [122] The developed doctrine of inertia, however, which had been clearly formulated by Descartes, appears nowhere in Heereboord's *Philosophia naturalis*. Even those troubling problems that were now at last resolved by the concept failed to evoke from Heereboord a reference to inertial motion. The physical arguments against the motion of the earth, such as the flying cannonballs and collapsing houses, were completely neglected, and no mention of inertia is to be found in Heereboord's treatment of the hoary problem of projectile motion. In his new commentary, Heereboord was inclined, ultimately, to attribute the continuing motion of projectiles partly to an impressed force and partly to the action of the air, as Aristotle had suggested.[123] Dealing, here, with a problem that derived from the failure to recognize the inertial character of motion, Heereboord's own continuing silence with regard to inertia was eloquent. His new commentary was, in fact, a retrogression, for in his original commentary, he had, like Burgersdijck, attributed projectile motion to the impressed force alone. As for the opinions of Descartes, *et al.*, Heereboord casually remarked elsewhere: "there is little difference between the received opinion and the opinion of those called the 'new philosophers'; hence, we shall not tarry among them." [124] That which Heereboord still chose to ignore in his later years, however, had been elaborated at Leiden for several years now by his younger colleague and former accomplice in the harassment of Stuart, Joannes de Raey.

When Heereboord turned his attention to the heavens, the philosophy of the *Philosophia naturalis* became no more adventuresome. The original scholastic thesis that the earth occupied the center of the world evoked no new thoughts from Heereboord himself, either pro or con.[125] Remaining noncommittal, he limited himself to brief references to the opinion of Regius and Descartes that the sun was the center of the world with the earth being carried around it by the movement of the rest of the heavens.[126]

Heereboord was also ambiguous with respect to the twofold division of the cosmos. In one of his newer commentaries, he upheld the division, stating that it was a division between two species, indeed, contraries, in the genus of natural bodies; their separation was not only one of dis-

[122] *Ibid.*, p. 33.
[123] *Ibid.*, p. 77.
[124] *Ibid.*, p. 83.
[125] *Ibid.*, p. 102.
[126] *Ibid.*, pp. 99 and 112.

tance, but of essence as well.[127] In another of the new commentaries, however, he chose to dissent from the thesis that the heavens were a fifth essence distinct from elements, but he did so, again, in a manner that had little in common with the new science.[128] The incorruptibility of the heavens, which he did not at this point reject, did not argue a fifth essence, he maintained, for the same matter and essence could be subjected to a form which was now dissoluble and now indissoluble (rather poorly put, since form and essence were closely identified in scholastic thought).

Further on, Heereboord did take up the question of the distinct and incorruptible heavens in the light of the astronomical discoveries which were now, however, hardly new. By the beginning of the 1660's, Heereboord's treatment of the question was surely a bit archaic. He spoke of "more noble philosophers" who, in opposition to the Aristotelians, considered the heavens subject to corruption and generation and even asserted that the four primary qualities associated with the elements were to be found in the stars and other celestial regions; "and this opinion is not absurd," opined Heereboord, "but indeed quite true, for experience itself bears witness to the corruption of the heavens and the generation of celestial bodies, and the telescope will teach us this as well." [129]

In yet another place, Heereboord described the heavens as consisting of the matter of fire or air, of which the stars were but condensations in an otherwise continuous body.[130] To support the similarity of the heavens and the air, he cited some rather peculiar reasoning. Recent philosophers, he wrote, considered the difference between the air and the heavens to be a difference of coarseness and density, "and they use this argument, which is not to be despised": if the heavens and the air were in fact different in substance and nature, then because of refraction (which neither he nor they apparently realized also followed from a change in density), no star would be seen in its proper place, and this, indeed, was absurd. Whether it was absurd or not, Descartes, for one, had asserted that this was precisely what occurred.

The Cartesian vision of the universe was also described in its broader outlines, but largely in special commentaries that, devoted specifically to summarizing the moderns, lacked any personal commitment on the part of Heereboord himself. The vortices and fluid heliocentrism of Descartes were briefly mentioned, as were the three Cartesian "elements,"

---

[127] *Ibid.*, pp. 92-3.
[128] *Ibid.*, pp. 93-4.
[129] *Ibid.*, p. 113.
[130] *Ibid.*, pp. 95-6.

and, though Heereboord had no doubt that the world was round and finite, the concept of a universe of "indefinite" size was defended as well.[131] In general, however, it appears that the more grandiose images of Descartes' universe – the vortices, the earth being rushed along in celestial currents, the grinding fragments of matter filling the heavens – failed to excite Heereboord's imagination.

It is apparent that with respect to physics, which Heereboord himself saw as embracing nearly the whole of Descartes' philosophy,[132] Heereboord could hardly be called a Cartesian, much less a spokesman for the new science. He rejected the fundamental Cartesian identification of matter and extension, and where the new philosophy of Descartes and the new science were most significantly joined, with respect, that is, to the new understanding of motion, he had made but little progress. The Cartesian imagery of particles of matter in motion, a passing but influential accoutrement of the new science, did arouse Heereboord's enthusiasm, but only in certain contexts, most notably where it offered an escape from the entangled scholastic cogitation regarding the peripatetic principles. His teachings on the cosmos and the heavens were confused and more cautious than Kyper's, and he made hardly any mention of mathematics and experimentation, keynotes of the new science.[133] He continued to evidence a passion for dialectic dispute, and his thought, despite his denigration of the scholastics,[134] still unfolded within the framework they had provided.

How, then, did Heereboord become the source of so much anxiety during the crisis over Cartesianism? The *Philosophia naturalis* suggests what he himself implied, that the stimulating disruptiveness of his presence at Leiden lay not in the modernity of his own philosophy but in his struggle to secure an open market place of ideas within the walls of the university.

It is not to be denied that, as I myself am freer for having been born under free Dutch skies, I granted my students freedom of wit and opinion in disputation. In

---

[131] *Ibid.*, pp. 98, 99, 111-2, 21-2, and 49-50.

[132] *Ibid.*, p. 1.

[133] He did cite a few "common experiments (*experientiae*)" relating to the question of the vacuum, but they were in fact very superficial observations probably deriving from the tradition of conventional and usually untried "experiments" frequently cited in this same context in the past (*Ibid.*, pp. 59-60; see Charles B. Schmitt, "Experimental Evidence for and against a Void: The Sixteenth-Century Arguments," *Isis*, Vol. LVIII (1967), pp. 352-66). He made no reference to the historic series of seventeenth-century experiments on the vacuum that began with Torricelli's work with the barometer.

[134] See for instance, "Ad curatores epistola," p. 6, and *Meletemata philosophica* [1654], p. 144.

those disputations of which I myself was author, I followed the principles of Aristotle, as is called for in the curricula of physics, ethics and "selected disputations." In the theses and corollaries, as they are called, prepared by the exertion and skill of the students themselves, I permitted the opinions and principles of other philosophers to be aired as well, so that their wits and mine together might be exercised and whither reasoning might lead us discovered.[135]

The hallmark of his own thought and career was not a dedication to any philosophic system, but a resistance to the subjection of philosophic speculation and debate to tradition and authority.

As was the case with so many of the intellectual dissidents of the time, it was also not Aristotle who was the object of Heereboord's hostility, but those, rather, who sought to establish the Aristotelian philosophy as inviolable authority, corrupting, even as they did so, the real burden of Aristotle's thought. Scholastic philosophy, declared Heereboord, was "enfeebled in many parts, entangled in many ways, detestable in the great barbarism of its speech, distended with many circumlocutions, erroneous in many opinions, replete with barren assertions," [136] but Aristotle himself remained for Heereboord an exemplar of the true philosopher, concerned with nature and not opinions, possessed, like Descartes, of a mind and spirit freed of preconceptions.[137]

Likewise, it was not the philosophical system elaborated by Descartes that had evoked Heereboord's lavish praises; it was the hope Descartes inspired that, after centuries of fettered constraint imposed by the scholastics, philosophy was at last reviving and on the verge of a new and decisive encounter with truth. Nor was Descartes alone in this liberation of philosophy, of which the new science was in fact but one manifestation. Among many others, there were as well Luis Vivés, Petrus Ramus – "that name and person odious to many and precious to many," – Francesco Patrizzi, Thomas Campanella, Bernardino Telesio, and the "illustrious heroes" Francis Bacon and Johannes Amos Comenius.[138] Nonetheless, it was Descartes, "the morning star of the dawning philosophy," [139] who was foremost among them. He had provided the key to true philosophy and had "opened the portal through which the unchanging truth of things is to be attained." [140] He could teach us to rid ourselves of uncertain opinions and preconceptions and to recover that

[135] "Ad curatores epistola," p. 9; see also pp. 10-1.
[136] *Meletemata philosophica* [1654], p. 144.
[137] "Ad curatores epistola," pp. 2 and 13.
[138] *Ibid.*, pp. 9-10; *Meletemata philosophica* [1654], pp. 334-5; "Consilium de ratione studendi philosophiae," included in the *Philosophia naturalis cum novis commentariis explicata.*
[139] "Consilium de ratione studendi philosophiae."
[140] *Ibid.*

"inestimable freedom of mind and judgement'" which Aristotle himself had once brought to philosophy.[141] It was the promise of a new golden age of philosophy that drew Heereboord to Descartes, and Heereboord's defiant espousal of his cause was only sharpened by the fact that Descartes himself now seemed the most prominent victim of the prejudice to which philosophy had so long been subjected.[142]

Despite his enthusiasm for the upheaval he saw taking place within philosophy, however, Heereboord failed to understand the most consequential component – or product, perhaps – of that upheaval, the new science. Francis Bacon had also enjoyed Heereboord's high regard, and in his inaugural address in 1641, Heereboord had urged in Baconian phrases a recourse to experiments, and he later attributed the dissension and uncertainty prevailing in natural philosophy to the disregard for experiments and the senses.[143] Nonetheless, experimentation played no role in the physics of the *Philosophia naturalis,* and Heereboord was being more faithful to his own temperament when he declared that the means to the acquisition of true knowledge in philosophy were lectures, reading, meditation, and disputation.[144] A recurrent theme stressed by Heereboord throughout his career was the need for the philosopher to turn from the study of men's opinions to the study of nature, but for Heereboord, the real significance of this counsel lay not in the apparent emphasis on a renewed observation of nature, but in the implicit rejection of any such "opinions" as authoritative in the practice of philosophy.[145]

Heereboord was, after all, a teacher of philosophy, devoted to exercising wits rather than probing the phenomena of nature. In the 1640's, his teaching had doubtless provided an invigorating antidote to the rigid traditionalism of an Adam Stuart, but in the 1650's when the chair in physics passed to the Cartesian Joannes de Raey, Heereboord's own natural philosophy, by contrast, remained deeply encumbered by tradition. His career argues that it was not an excessive deference to an acknowledged authority in philosophy, nor fear of novelty, nor anxiety over administrative disapproval that most obstructed the fuller acceptance of the methodology and growing body of knowledge encompassed by the new science; it was rather the continuing efficacy of deeply ingrained philosophic and pedagogical preconceptions, which Heereboord himself rightly recognized as the greatest barrier to philosophic freedom.

[141] "Ad curatores epistola," p. 13.
[142] See Heereboord's remarks on the reaction to his address, "De libertate philosophandi"; *Ibid.*
[143] *Ibid.*, p. 6; *Meletemata philosophica* [1654], p. 145.
[144] "Ad curatores epistola," pp. 9-10.
[145] *Ibid., passim.*

CHAPTER IV

JOANNES DE RAEY:
THE INTRODUCTION OF CARTESIAN PHYSICS
AT LEIDEN

Despite their continuing prohibition against the public mention of Descartes' name or philosophy within the university, the curators at Leiden were far less hostile to the new philosophy than the prohibition alone would seem to suggest.[1] The philosophical and theological issues that so troubled the professorial combatants were not in themselves the curators' first concern. It was the turbulent unrest within the university, and not Cartesianism, that they sought to suppress. The States of Holland and West Friesland, likewise, were not attempting to dictate in matters of philosophy in their decree of 1656. In calling for the abandonment of those Cartesian theses that some persons, at the time, found offensive, the States were seeking to quiet passions rather than to establish an official position in philosophy.[2]

That the discord within the university would be a source of such concern is understandable, for such internal conflict not only invited outside intervention – of which the decree of the States was itself an example – but also threatened the prestige of the faculty and, hence, the prosperity of the university.[3] In an age so characterized by sporadic student violence, moreover, the intense student involvement posed other dangers as well. In 1659, the States of Holland and West Friesland felt compelled to follow their decree against disruption in the lecture halls with a further decree against the carrying of pistols and other weapons by members of the university, a student vogue against which the university administration

---

[1] Sassen, "Adriaan Heereboord," pp. 16-7, and *Geschiedenis van de Wijsbegeerte in Nederland*, p. 151. Dibon, "Notes bibliographiques sur les cartésiens hollandais," in *Descartes et le cartésianisme hollandais*, E. J. Dijksterhuis, *et al.* (Paris: Presses Universitaires de France; Amsterdam: Editions Françaises d'Amsterdam, 1950), pp. 266-7.
[2] *Bronnen*, Vol. III, p. 57*. Thijssen-Schoute, "Le Cartésianisme aux Pays-Bas," p. 210.
[3] de Vrankrijker, *Vier Eeuwen*, pp. 39 and 52.

had been struggling for over half a century.[4] As late as 1682, the curators were trying to prevent the repetition of a recent armed attack on the municipal night watch.[5]

While persistently seeking to allay the turbulence within the university, however, the curators displayed a degree of nonpartisanship in philosophical matters that, despite the formal prohibition, served as much to shelter as to obstruct the progress of Cartesian thought at Leiden. Even the troublesome Heereboord appears to have enjoyed no less, and probably more, favor among the curators than did Adam Stuart.[6] In the context of other evidence of administrative sympathy with the Cartesian camp, the prohibition itself may be looked upon as essentially an effort to induce the proponents of the new philosophy to urge their case with greater tact. If the latter would abstain from pointed references to Descartes as their source of inspiration, if they would avoid those inflammatory questions that were only too well known, they could expect a policy of benevolent neutrality on the part of the curators.

Though their offer went too often unheeded, the curators pursued such a policy of benevolent neutrality nonetheless, a neutrality, moreover, that was more than merely passive. They proved willing to grant the Cartesians a voice within the faculty, and in 1653, only six years after they had first declared their prohibition ostensibly against Descartes, the curators transferred the chair in physics from Stuart to a known Cartesian with a record of activism dating back to his student days under Regius at Utrecht.[7] It was, indeed, the same Joannes de Raey whom Stuart himself had singled out as Heereboord's accomplice in 1648.[8]

De Raey had come to Leiden from Utrecht to complete his studies and in 1647 had received both his masters in philosophy – under Heereboord – and his doctorate in medicine. Without the consent of the academic senate, he began to offer private lectures on Cartesian metaphysics, which soon elicited a letter of complaint from Jacobus Revius in 1648.[9] The academic senate, in response, ordered de Raey to abandon these private

---

[4] *Bronnen*, Vol. III, p. 115*; see also Vol. I, pp. 291* and 311*-4*, and Siegenbeek, *Geschiedenis der Leidsche Hoogeschool*, Vol. I, p. 172.
[5] *Bronnen*, Vol. IV, pp. 1*-2*.
[6] Sassen, "Adriaan Heereboord," *passim*.
[7] *Bronnen*, Vol. III, p. 76. Haldane, *Descartes*, p. 235.
[8] See above, pp. 45-6. On de Raey, see, in addition to general works on Cartesianism in the United Provinces, G. C. B. Suringar's "Invloed der Cartesiaansche Wijsbegeerte op het natuur- en genees-kundig Onderwijs aan de Leidsche Hoogeschool," in his *Bijdragen tot de Geschiedenis van het Geneeskundig Onderwijs aan de Leidsche Hoogeschool*.
[9] *Bronnen*, Vol. III, pp. 14*-5*.

lectures and to avoid the Cartesian philosophy.[10] In 1651, however, the curators granted him the right to teach the *Problemata* of Aristotle, though on the condition, according to a now familiar formula, "that he always remain within the limits of the Aristotelian philosophy, which alone is accepted within this university . . . ." [11] Two years later, he was appointed professor of physics and, further, in 1658, was allowed to teach the *Institutiones medicinae* as well, a duty, however, from which he was later dismissed at his own request following difficulties with the medical faculty.

De Raey was later described by a Silesian theology student at Leiden as a true Dutchman in speech as well as manners, caring little for courteousness, and de Raey's relations with his colleagues on the faculty were not free of angry outbursts.[12] Nonetheless, he did not pursue direct confrontation as relentlessly as did Heereboord, nor did he cultivate such bitter acrimony. Unlike Heereboord, de Raey was himself clearly a Cartesian, but he attempted to represent Descartes' philosophy as compatible with Aristotelian thought, more compatible, indeed, than the corrupted legacy of the scholastics. The curators had been informed by Golius, the rector of the university in 1652, that de Raey avoided whatever might cause offense in his disputations and taught the philosophy of Aristotle, but in such a way as to distinguish that philosophy from the "monkish philosophy." [13]

In arguing the similarity of Cartesian and Aristotelian thought, de Raey associated himself with a broader seventeenth-century effort to reconcile the two philosophic traditions.[14] But de Raey himself remained first and foremost a propagandist for Cartesianism, and how deeply convinced he really was of the compatibility of the two philosophic traditions is open to question. In the dedicatory letter of his *Clavis philosophiae naturalis, seu introductio ad naturae contemplationem, Aristotelico-Cartesiana*, published the year after he had assumed the chair in physics, de Raey acknowledged that the primary reason for his reconsideration of Aristotle was the latter's continuing hold on the university. Aristotle's role in the "Aristotelian-Cartesian" exposition of Cartesian philosophy that followed was that of a forerunner who had intimated

---

[10] *Ibid.*, p. 11.
[11] *Ibid.*, p. 54.
[12] Schotel, *De Academie te Leiden*, p. 235. Cramer, *Abraham Heidanus*, p. 76.
[13] *Bronnen*, Vol. III, p. 60.
[14] See Bohatec, *Die cartesianische Scholastik*. It was suggested that de Raey influenced similar efforts by Leibniz, but the latter himself denied this (Lewis White Beck, *Early German Philosophy: Kant and His Predecessors* [Cambridge, Mass.: The Belknap Press of Harvard University Press, 1969], p. 182, n. 22).

what was more clearly and completely realized by a distant successor. It was a one-way dialogue; where disagreement was acknowledged by de Raey, it was Aristotle who had erred. De Raey communicated no deep belief in the fundamental harmony of the two philosophic systems, but argued, rather, that there were significant points of agreement which could serve as the basis for a philosophy more certain and more reliable than the corrupted Aristotelianism of the scholastics. This more reliable philosophy, however, proved to be Cartesianism itself.

The *Clavis philosophiae naturalis,* the publication of which disregarded the explicit prohibition against books about Cartesianism, was dedicated to the curators of the university themselves, who first agreed to grant de Raey an honorarium in recognition of the dedication, but then reconsidered and decided to withhold the honorarium because of the reference to Descartes in the title.[15] The work does not appear to be a direct duplication of de Raey's lectures or disputations. The broad format conforming to the organization of a whole course in physics, such as had characterized the texts of Burgersdijck, Kyper and Heereboord, is missing, as is the pervasive influence of the disputation. De Raey was writing for a wider audience than students alone, and the natural philosophy he put forward pressed its claim to credibility by an appeal to mental imagery rather than to dialectic.

De Raey concentrated in the *Clavis philosophiae naturalis* on what was, in effect, the problem of philosophic first principles, though he wrote rather of *praecognita,* those things which were to be known and understood first and from which all other knowledge was to follow. He posited four such *praecognita* in physics: the essence of matter, the origin of motion, the communication (or alteration) of motion, and the nature of an extremely subtle and mobile form of matter that permeated the coarser bodies of the sublunar world.[16] What followed in the *Clavis philosophiae naturalis* was an explication of the Cartesian principles of matter in motion. The inclusion of the fourth *praecognitum,* the subtle matter, reflected a more personal emphasis in de Raey's philosophy, but derived no less from the world picture of Descartes.

In his chapter "Concerning the nature of body, or the matter which philosophers call first matter," de Raey began indeed with the texts of Aristotle, only to conclude, however, with the Cartesian identification of matter and extension. Having cited Aristotle at length, de Raey at last concluded that the matter of which Aristotle spoke – which many, he said, would consider nothing at all – was in fact nothing other than

[15] *Bronnen,* Vol. III, pp. 52-3, 93 and 107.
[16] *Clavis philosophiae naturalis,* pp. 47-8.

extended being (*res extensa*), "that very same indefinite extension which in particular bodies is determined by a fixed quantity, length, breadth and thickness (*crassities*) . . . ." [17] Heereboord had pointedly maintained that it was best only to say that matter was extended, but for de Raey, matter was now pure extension itself, a conceptual shift the significance of which lay in what was shorn away. Whatever exceeded the perceivable content of pure spatial extension was now alien to the content of matter as well, and with its content thus emptied, matter was stripped of all possibility of life and awareness and purged of the innate potencies and internal principles which both scholastic and Renaissance philosophers had generously strewn throughout the whole of nature.

This feeling for the emptiness of matter was conveyed most vividly in the Cartesian argument against the void. Since the extension to be found in matter, Aristotle argued for de Raey, is no different than the extension of the void, no more (or less) reality was to be ascribed to the extension of one than to the extension of the other.[18] The conclusion to be drawn, which de Raey did not, however, develop in the *Clavis philosophiae naturalis*, was the impossibility of the void, which, by its very dimensions, was itself rendered matter. Despite its emptiness, extension was nonetheless a complete and total substance – "if this is not substance," Aristotle was compelled to testify, "whatever else is escapes us" [19] – the only substance, in fact, as a student of de Raey's later stressed, which could possibly be considered to exist in natural bodies.[20]

By reducing the substance of natural bodies to pure extension, a crucial step was taken in the reconceptualization of motion as well, for the emptiness of natural bodies now deprived them of the scholastic internal cause or principle of movement. What could be more evident, challenged de Raey, than the irrelevance of motion to the essence of matter.[21] "What is clearer to a mind free of preconceptions (*praejudicia*) than the absence of any virtue or tendency sufficient for self-movement in the corporeal mass (*moles*) of extension which alone constitutes the essence of matter?"

The classical atomists, he continued, had erred by endowing matter with an intrinsic and eternal source of motion that accounted, without

---

[17] *Ibid.*, p. 53.
[18] *Ibid.*, pp. 53-4.
[19] *Ibid.*, p. 53.
[20] Henricus vanden Velden, *Disputatio de substantia, pars secunda*, sub praesidio D. Johannis de Raei (Lugduni Batavorum: Abrahamus à Geerevliet, 1659), VIII.
[21] *Clavis philosophiae naturalis*, p. 68.

God, even for the creation of the world.[22] The common multitude of men, moreover, had also gone far astray in attributing to natural bodies not only an internal principle of motion, but a principle frequently possessed of desire and intelligence bent on the achievement of some end. In contrast to these, asserted de Raey, Aristotle had rightly declared that no body could in fact be the efficient cause of its own motion. Despite de Raey's real approbation of this Aristotelian maxim, however, the true victim of the new doctrine of the barrenness of matter was, after all, Aristotle's own doctrine of natural motion.

De Raey was not so courteous towards his venerable ally as to free him of criticism altogether, and he took him to task for his attribution of perpetual motion and divinity to the heavens.[23] In truth, said de Raey, the heavenly bodies are but immense, brute lumps of extension, and it follows from the teachings of Aristotle himself that they are moved not by their own but by an alien power, a power of such immensity that it could only derive from God, the true cause and origin of all motion. Aristotle's error, however, was to be ascribed less to the man than to the darkness of the age in which he lived, not to his philosophy, but to the paganism and beliefs of his homeland. It in no way diminished the truth and fruitfulness of his axiom that nothing could be the source of its own motion, which de Raey now chose, for better or for worse, as the means by which to guide his reader to the concept of inertia.

In the opening pages of his chapter on "the communication of motions and the interactions of bodies," de Raey proposed that his reader consider what happened to a single isolated body placed either in a state of motion or rest. It followed from its inability to move itself, observed de Raey, that such a body at rest would remain at rest and tend towards no other place until expelled from its present place by some external impulse (*impetus*). Likewise, there was no reason why a body, once moving, should not perpetually continue at the same speed if it met nothing to diminish or arrest its impetus (*impetus*). In effect, de Raey had asked his reader to abandon the conviction that every motion had a natural end and to step into a world of inertial motion.

Heereboord testified to the difficulty of this step by his continuing refusal to attempt it, and de Raey, in stressing Aristotle's maxim, may, in the final analysis, have provided as much an obstacle as an aid. What, in truth, might seem a greater contradiction of the assertion that nothing moves itself than the perpetual movement of an isolated body? For de Raey, as for Descartes, all motion had an unceasing and perpetually

[22] *Ibid.*, pp. 69 ff., 85 and 86.
[23] *Ibid.*, p. 77 ff.

immediate cause in God, and de Raey's treatment of motion in the *Clavis philosophiae naturalis* is highly charged with religious sentiment. To what extent a continuing divine causality actually suggested physical cause to de Raey is uncertain, but his recourse to Aristotle's maxim as an introduction to the concept of inertia bespeaks in itself his own failure to have grasped as yet the fundamental novelty in the new conception of motion, that the perseverance of motion no longer required any natural cause at all.

Continuing further, de Raey asked his reader to conceive of a single, undivided body. Once set spinning about its center, said de Raey, this body would always continue spinning if there were nothing surrounding it to impede it. Or if it were moved in a straight line through a space where it encountered no resistance, no one – as Aristotle was forced again to testify [24] – could say by what reason the body would stop at one place rather than any other. With his image of the spinning body, however, de Raey had obscured the purely rectilinear concept of inertia formulated in Descartes' second law of nature, and neglecting this second law, de Raey concluded his exposition of inertia with the more general first law alone, "that, to be sure, *every body, to the extent that it is simple and undivided, persists, in so far as it can, in the same state and is never altered unless by external causes."* [25]

Having progressed so far in the new perception of motion, de Raey appears to have faltered before what proved its most baffling feature, its relativity, wherein lay the final recognition of persisting motion's independence of causality. An important feature of Descartes' doctrine of motion in the *Principia philosophiae*, the relativity of motion was hardly mentioned in the *Clavis philosophiae naturalis*; for de Raey, the concept was surely a source of deep uneasiness. Once motion had become a mere change in the relative position of neighboring bodies, how could the single moving bodies de Raey had asked his reader to conceive be conceived to move at all? In the mind's eye, the isolated body seemed to have lost the very possibility of motion and the idea of motion itself to have become meaningless. If de Raey found this paradoxical situation unacceptable, he only shared the perplexity of his age. Even Descartes had not been able to retain his grasp of his own elusive concept, and de Raey's omission is all the more conspicuous in the light of his emphasis on two other Cartesian doctrines which were themselves also incompatible

---

[24] Aristotle is cited to this effect from his argument against the void (*Physics*, IV, 8), where he obviously expected the situation he was describing to be considered unacceptable.
[25] *Clavis philosophiae naturalis*, p. 107.

with the relativity of motion, the permanent quantity of total motion in the world and the laws of impact. The latter, indeed, constituted de Raey's third *praecognitum*.

The importance of the laws of impact, elaborately explained and illustrated by de Raey, followed naturally in the mechanical universe of the new philosophy, where all the phenomena of nature arose only from the unceasing collision of parts of matter in motion. Excepting the first, however, the seven Cartesian laws of impact were wrong, and an initial source of error was the failure to appreciate the implications of the relativity of motion. Within the context of relative motion, the collision of two bodies is always the same single type of event, but the Cartesian laws of impact conceived at least three distinct types of collision: one body initially being at rest in one case, both bodies moving towards each other in another, and both bodies moving in the same direction with a faster body striking a slower from behind in a third. Failing to recognize, as well, that both of two unequal bodies, the larger as well as the smaller, must necessarily be affected when they collide, the laws elaborated eight unique cases of impact between two bodies (to which de Raey added a ninth of his own) with a different pattern of mutual effect and reaction in nearly every case. In complete disregard for the significance of relativity, rest was conceived throughout as an absolute state fundamentally distinct from motion; ironically, however, there was no way within the framework of these laws whereby this state could ever be restored to a body once it had been disturbed.

Regardless of the absence of relativity, however, de Raey had described in his physics a motion conceived now as a precise and measurable quantity – the product of the quantity of matter and its velocity – and conforming to equally precise and invariable laws, even if, for the moment, incorrect. Motion had been freed of the influence of ends or goals in nature, and though bodies continued, indeed, to conform to an order in nature – a most beautiful order, without confusion or mishap, said de Raey – it was an order, he stressed, of which these bodies had no knowledge or awareness.[26] He had left the principles of nature, now matter and motion, bearing no vestige of order within themselves beyond a consistent and measurable perseverance; to this extent, de Raey had embodied in his phyics the reconceptualization of nature fashioned by the new science of the seventeenth century.

Even these barren principles of extension and persistent motion, however, when freed from a rigorous subordination to mathematics and

[26] *Ibid.*, p. 108 ff.

strict empirical verification, were susceptible to whimsical flights of the imagination, and Cartesian natural philosophy was characterized by the extravagant world picture elaborated by Descartes in the *Principia philosophiae* no less than by its progress in the formulation of the concepts of modern science. In his first chapter on motion, de Raey had sketched the general outlines of Descartes' image of celestial space, a vast ocean of fluid matter swirling in huge and innumerable vortices, each with a star at its center.[27] The sun was also such a star, and the celestial currents circling about it bore with them the dense, hard bodies of the planets, including the earth. De Raey himself showed no hesitation in finally asserting the heliocentrism of the solar system,[28] but he reaffirmed, nevertheless, the validity of Descartes' own evasion – that the earth, being motionless with respect to the fluid heavens that moved it, could justly be said to be at rest – and in so doing, provided one of the few echoes in the *Clavis philosophiae naturalis* of the relativity of motion.

De Raey's description of the Cartesian heavens was far too brief to indicate to what extent he accepted or understood Descartes' luxuriant celestial elaborations. Like Heereboord, he was captivated far more by the Cartesian imagery of a world of shaped and moving parts of matter so small as to lie beyond the reach of the senses. The observable world became an outer shell masking a whole new unseen realm of nature that lay below it everywhere: "Or is it not with bodies created by God," he asked, "as it is with buildings made by men, that more skill and beauty lie concealed within than are seen on the surface when we first arrive?" [29] The discovery and exposure of these inner secrets ranked among the first duties of the natural philosopher.

The ordinary man observes only the external variations and forms of things; the philosopher must descend into the interior, reveal the structure, nexus and intervening spaces of the tiniest parts, observe hidden movements and fully expose, at last, the ingenious art of nature together with its causes.[30]

The stress on the imperceptibly small structure and workings of nature found expression in de Raey's selection of the Cartesian "subtle matter" as the fourth *praecognitum* of his natural philosophy. Not so universal in its significance as the other *praecognita*, it was nonetheless necessary, maintained de Raey, for the understanding of particular phenomena,

---

[27] *Ibid.*, p. 62 ff.
[28] *Ibid.*; see also p. 22.
[29] *Ibid.*, p. 26.
[30] *Ibid.*, p. 32.

especially those observed about the earth.[31] It consisted, indeed, of that same fluid matter which filled the heavens; pressing closely about the earth, it permeated terrestrial bodies and gave rise to those phenomena in nature which amazed the common multitude and had until this time defied all explanation. A close reading of the several works of Aristotle showed, at least to de Raey's satisfaction, that Aristotle himself had also recognized the presence of such a celestial matter mingling with the bodies of the sublunar world.

This matter was composed of the first two of the three Cartesian "elements," the three principal forms of matter resulting from the fragmentation of world extension by God's original creation of motion.[32] Extension had been shattered into a vast ocean of milling particles whose constant mutual abrasion had inevitably reduced them to minute, imperceptible spheres, possessed, by reason of their size and shape, of an extreme agitation and mobility. In the very process of wearing each other down, however, they had produced a still finer form of matter, a residual sawdust, as it were, that flowed continually between the spheres themselves. So fine and fluid was this dust that nature had no nook or cranny too small for it to fill. Together with the spheres, this dust composed the subtle matter which permeated the grosser bodies of the earth.

To illustrate this penetration, de Raey suggested that his reader consider earthly bodies to have been originally composed from many smaller corpuscles of varying shapes and sizes. None of these corpuscles, he explained with numerous drawings, could have so perfectly come together as to have avoided an innumerable quantity of gaps and fissures, and it was the presence of the hyperactive celestial matter in just such gaps in earthly bodies that caused the multitude of previously inexplicable natural phenomena. It accounted for the expansiveness of heated air, the fluidity of liquids, the effects of light and even the revival of life that followed in springtime after the cold sluggishness of winter.[33] Such varied and previously mystifying phenomena as magnetism, rarefaction and condensation, fermentation, and barometric experiments now all attested to the constant passage of this subtle matter through the internal spaces of terrestrial bodies.

It was an intriguing vision of ceaseless activity in the submicroscopic innards of nature; beyond merely gratifying the imagination, however, and despite the excesses to which it was prone, this fantasy mechanized in its own crude fashion the most obscure processes of nature. The prin-

[31] *Ibid.*, p. 126 ff.
[32] *Ibid.*, p. 142 ff.
[33] *Ibid.*, p. 190 ff.

ciple of form was gone, and the qualities and properties that once bespoke its presence had been reduced to unseen particles of matter in motion. A physical world once interpreted in terms derived from the effort to conceptualize the acquisition and loss of being and identity had now been rendered into an imagery so easy to conceive that it raised the highest hopes for the ultimate attainment of a total understanding of nature.

The natural philosophy of the *Clavis philosophiae naturalis* remained, however, still a poor reflection of the complex and still-evolving synthesis of ideas and activities which was the new science. Like so many under the spell of Cartesianism – following, indeed, the example of Descartes himself in the *Principia philosophiae* – de Raey failed to give due regard to the procedures being used and refined in the course of the scientific advance and was too readily satisfied with the efforts and pleasures of the imagination alone. From top to bottom, from the stellar vortices to the most subtle sense perceptions, nature had been mechanized, but it lacked as yet the mathematical analysis through which the mechanization of nature was to lead to the great achievements of modern science. All things in de Raey's world were now measurable, but little was measured. Despite a display of obvious enthusiasm for the mercury tube of Torricelli,[34] de Raey also paid little heed to the role of experimentation. Throughout his natural philosophy, it remained the compelling vividness of the Cartesian imagery which determined his understanding of natural phenomena.

Regardless of its shortcomings with respect to the new science, nonetheless, the Cartesian system which de Raey so vigorously put forward offered to restore the internal cohesion that had long been disintegrating in the physics of the schools. The Cartesian world was not without its internal strains, from the inconsistencies of its intricate elaborations to the difficulty of maintaining pure extension as the frangible, solid substance whose fragments and movements composed the world. For a generation brought up, however, against the background of a traditional physics whose cosmos was rapidly collapsing and whose principles appeared increasingly irrelevant to active scientific enterprise, such strains might well seem petty, hardly worth jeopardizing the new coherence otherwise offered by the unifying imagery of matter in motion.

Within the framework of this new coherence was completed the disintegration of the old. Two concepts, the cosmos and the natural body, had been keystones in the order which peripatetic natural philosophy had imposed on the physical world; one of these, the cosmos, had passed away

---

[34] *Ibid.*, p. 191 ff.

completely. The last remnants of the celestial spheres had been dissolved in the Cartesian vortices, and the center of the cosmos, about which all its parts had been disposed, was lost when the earth was moved about the sun and the sun itself became but one among many stars. What remained was a universe of undifferentiated space where nature and its principles recognized no direction or place. A last lingering vestige of the cosmos, which had embraced an immovable space, could perhaps be detected only in the persisting, if now anomalous, idea of the absolute nature of rest.

The drama of the Cartesian imagery obscured the degree to which the peripatetic concept of the natural body had suffered as well. A distinctive but transient identity, the natural body had been the substance around which scholastic physics had revolved. Guided by brute perseverance alone, however, the mechanical principles of matter and motion were totally indifferent to identities that might appear and pass away in nature. The natural philosopher was no longer preoccupied with the coming-to-be and passing-away of an ever-changing multitude of individual substances, but with the ever-changing visage of a single persisting substance – matter – in motion. A disputation defended under de Raey in 1668, his last year at Leiden, had begun with a redefinition of physics:

Physics is commonly defined as *the science of the natural body in so far as it is natural*. But since the meaning of "nature" and "natural" remain to this day obscure, a far clearer definition would be, *the science of the visible world,* for the task of physics is none other than the explanation of the sensible phenomena of the world through intelligible causes.[35]

The natural identity as a pivotal concept in physical thought was disintegrating and only the phenomena, random manifestations of a universal and indifferent mechanical activity, would remain.

---

[35] Abrahamus de Reus, *Disputatio philosophica, de constitutione physicae,* sub praesidio D. Johannis de Raei (Lugduni Batavorum: Apud Viduam et Haeredes Johannis Elsevirii, 1668), I.

CHAPTER V

PASSING CRISES, ENDURING DISAGREEMENT

During de Raey's professorship at Leiden, the university emerged as an important center for the dispersion of Cartesianism throughout northern Europe, a role that was further enhanced in the 1660's by the presence at Leiden of Arnold Geulincx.[1] A convert from Catholicism, Geulincx had come to Leiden from Louvain in 1658.[2] Supported by the theologian Abraham Heidanus, he was appointed lector in logic in 1662 and three years later acquired the title of professor. Although he held that title ostensibly with the understanding that he would adhere to the peripatetic philosophy,[3] he was a dedicated Cartesian whose writings were to have a considerable impact in the following decades. The subjects for which he was formally responsible at Leiden were logic and ethics,[4] outside of which his major influence was in the realm of metaphysics. Nonetheless, physics was neglected neither in his writings nor teaching, and he bolstered the dominance that de Raey had already acquired for Cartesian natural philosophy within the university.[5]

In 1668, however, de Raey left the university for the Athenaeum in Amsterdam, and the Englishman Samuel Colepresse wrote to Henry

---

[1] De Raey's former students were particularly influential in the spread of Cartesianism in Germany. See Heinz Schneppen, *Niederländische Universitäten und Deutsches Geistesleben, von der Gründung der Universität Leiden bis ins späte 18. Jahrhundert* (Münster, Westfalen: Aschendorff, 1960), pp. 76-8. Bohatec, *Die cartesianische Scholastik*, pp. 51 and 78. Beck, *Early German Philosophy*, pp. 183-4.

[2] On Geulincx, see J. P. N. Land, *Arnold Geulincx und seine Philosophie* (Haag: Martinus Nijhoff, 1895); "Arnold Geulincx te Leiden (1658-1669)," *Mededeelingen, Kon. Akademie van Wetenschappen, Amsterdam, Afdeeling Letterkunde*, series 3, Vol. III (1887), pp. 277-327; "Arnold Geulincx and His Works," *Mind*, Vol. XVI, pp. 223-42. Victor Vander Haeghen, *Geulincx: Étude sur sa vie, sa philosophie et ses ouvrages* (Gand: Ad. Hoste, 1886).

[3] Land, "Geulincx te Leiden," p. 315.

[4] *Bronnen*, Vol. III, pp. 179* and 210*; see also the *Series Lectionum* for the years 1663-9. Land, *op. cit.*, p. 318.

[5] See his *Sämtliche Schriften in fünf Bänden*, ed. H. J. de Vleeschauwer (Stuttgart-Bad Cannstatt: Friedrich Frommann, 1965-68).

Oldenburg that with the departure of "ye famous De Raei," Leiden was likely to return to Aristotelianism again.[6] The future of Cartesianism at Leiden looked all the gloomier when, in the following year, Geulincx fell to the plague. But the curators again gave evidence of their willingness to cultivate the new philosophy by more than compensating for its recent losses. In 1670, they filled out the faculty of philosophy by the appointment of Theodorus Craanen and Burchardus de Volder, both, like de Raey, medical men and known Cartesians.[7] De Volder, who had acquired his doctorate of philosophy, or master of arts, at Utrecht in 1660 and his doctorate in medicine at Leiden in 1664, was also a former Mennonite who had to be baptized and received into the Reformed Church before assuming his post.[8] The number of Cartesian sympathizers was further augmented the following year by Christophorus Wittichius, who joined Heidanus, himself a former friend of Descartes and long-time defender of his philosophy, on the hypersensitive faculty of theology.[9] If anything, the balance now had shifted in favor of the representatives of Cartesianism, the less restrained of whom were soon to launch another aggressive campaign against the continuing defenders of the scholastic tradition. A last confrontation lay, indeed, in the very near future, in large part the consequence of international developments that were already in progress.

In the spring of 1672, Louis XIV of France launched the invasion of the Dutch Republic which he had long been preparing.[10] A large and highly-groomed French army rapidly penetrated deep within the Dutch borders, and the republic found itself suddenly confronted with the prospect of total ruin. In the panic that swept the provinces, the regime

---

[6] *The Correspondence of Henry Oldenburg*, ed. and trans. A. Rupert Hall and Marie Boas Hall (Madison, etc.: University of Wisconsin Press, 1965), Vol. V, p. 141.

[7] *Bronnen*, Vol. III, pp. 236 and 245. Sassen, *Geschiedenis van de Wijsbegeerte in Nederland*, p. 159.

[8] Suringar, "De Leidsche Hoogleeraren in de natuurkundige Wetenschappen, inzonderheid in de Kruid- en Scheikunde, na den Dood van Sylvius en vóór Boerhaave's Benoeming tot Professor Chemiae (1672-1718)," in *Bijdragen tot de Geschiedenis van het Geneeskundig Onderwijs aan de Leidsche Hoogeschool*, p. 18.

At Leiden, as at other continental universities, the degree of doctor of philosophy appeared as an equivalent to the master of arts in the course of the seventeenth century; Morison, *The Founding of Harvard College*, p. 145, n. 5.

[9] See Bohatec, *Die cartesianische Scholastik, passim*, and Cramer, *Abraham Heidanus*.

[10] See John B. Wolf, *Louis XIV* (New York: W. W. Norton and Company, Inc., c. 1968), p. 213 ff., and Pieter Geyl, *The Netherlands in the Seventeenth Century, Part Two 1648-1715* (London: Ernest Benn Ltd.; New York: Barnes and Noble, Inc., c. 1964), p. 121 ff.

of the oligarchic regent class collapsed; its leading statesman, Johan de Witt, was dismembered by the mob in the streets of The Hague, and William III, Prince of Orange, was restored to the quasi-monarchic powers that his ancestors had enjoyed but which had been denied him since childhood by the oligarchs of Holland. In June, the French advance was finally blocked by opening the dikes and inundating the countryside, but Dutch society had been deeply shaken.[11]

The war continued for six more years, during which time William III exercised extensive extraordinary powers throughout much of the republic. As in 1619, the conservative wing of the Reformed Church had closely allied itself with the cause of the House of Orange and, hence, once again briefly benefitted from the prince's political success. Among the consequences of the disaster of 1672, therefore, was a resurgence of the forces of intellectual reaction that soon made itself felt within the universities. Cartesianism would again become a major source of acrimony, particularly as a result of the efforts of Wittichius and Heidanus to fashion out of Cartesian thought a new framework for the elaboration of Reformed dogma.[12] Its opponents would further associate Cartesianism with the unsettling theological influence of yet another professor of theology at Leiden, Johannes Coccejus, who, like Geulincx, had died in the plague of 1669.[13] In the drastically altered political circumstances, the curators at Leiden now found themselves between resurgent conservative ecclesiastics on the one hand and the increasingly aggressive Cartesians on the other, the latter having been but recently strengthened within the university by the curators themselves.

In the summer of 1673, the issue of Cartesian doubt reappeared in accusations brought against Theodorus Craanen. As subregent of the college for theology students, Craanen had earlier that year been instructed to teach there only the logic and metaphysics of Burgersdijck, but the theologian Frederick Spanheim now complained to the curators of certain theses defended under Craanen on the propriety of doubting the existence of God.[14] Observing that Craanen, who had also been teaching Cartesian physics and physiology, was not likely to teach the peripatetic philosophy with the necessary enthusiasm, the curators

[11] Directly in the path of the French advance, Leiden had distinguished itself among the other cities of the province of Holland by its willingness to accept the humiliating French terms for surrender; Stephen B. Baxter, *William III and the Defense of European Liberty, 1650-1702* (New York: Harcourt, Brace and World, Inc., c. 1966), pp. 75 and 77.
[12] See Bohatec, *Die cartesianische Scholastik*.
[13] Sassen, *Geschiedenis van de Wijsbegeerte in Nederland*, pp. 141-2. Cramer, *Abraham Heidanus*, p. 120; but see also pp. 43 and 76.
[14] *Bronnen*, Vol. III, pp. 271 and 274-8.

removed him from the faculty of philosophy to that of medicine before the year was out.[15]

Craanen was also replaced as subregent of the theological college, and the brief career at Leiden of his successor there, Gerardus de Vries, testified to the reviving bitterness within the university. Appointed in December of 1673, de Vries informed the curators in the following June of his desire to depart.[16] The curators attempted to dissuade him, but he explained that, since the beginning, he had suffered abuse from those who sought with "indiscreet zeal" to suppress the old peripatetic philosophy. He now hoped to go where he could carry on his teaching with greater benefit and less ill-will. Angered by the harassment and classroom disturbances that were driving de Vries away – and Craanen appeared to have connived in these – the curators unanimously resolved to defend the peripatetic philosophy and maintain the States' resolution of 1656, which the curators felt had been scornfully disregarded.[17] The members of the faculties of theology and philosophy were called in, and the latter were "sharply" forbidden to abuse, censure, or impugn, directly or indirectly, in disputations, writings, or by any other means, the peripatetic philosophy or those who supported it. They were to have care that they stayed within their proper subject and to conform thereby to the earlier resolution of the States. To further bolster the hard-pressed defenders of Aristotelianism, Wolferdus Senguerdius and Wilhelmus Wilhelmius, both of whom had acquired reputations for their attachment to the traditional philosophy, were appointed professors of philosophy in 1675 and 1676.[18] The vigorous onslaught of the Cartesians continued nonetheless, and both Senguerdius and Wilhelmius now suffered the same student harassment and abuse that had driven de Vries away (and which contributed perhaps to Wilhelmius' own death in 1677, just a year after his appointment.) [19] The alarmed curators now spoke of a "general conspiracy" that was determined to stifle the peripatetic philosophy at Leiden and establish the dominance of Cartesianism.[20]

To be more fully understood, the alarm of the curators must also be seen against the background of the growing restiveness within ecclesiastical circles outside the university. During these same years, conservative

---

[15] Suringar, "Invloed der Cartesiaansche Wijsbegeerte op het natuur- en geneeskundig Onderwijs aan de Leidsche hoogeschool," p. 29.
[16] *Bronnen,* Vol. III, pp. 278 and 290-1.
[17] *Ibid.,* pp. 280 and 291-3.
[18] *Ibid.,* pp. 307, 319, 321, and 259*-60*. Sassen, *Geschiedenis van de Wijsbegeerte in Nederland,* p. 160. Concerning Wilhelmius, see also the *Nieuw Nederlandsch Biografisch Woordenboek,* Vol. V, 1125-6.
[19] *Bronnen,* Vol. III, pp. 314-5.
[20] *Ibid.*

churchmen, enjoying the favor of the Prince of Orange, had mounted their own campaign against "pernicious novelties" imputed particularly to the influence of Coccejus and Descartes. This campaign finally came to bear directly on the university in late 1675, when the ecclesiastical organization for Walcheren, in Zeeland, following the initiative of colleagues at der Goes, submitted to the curators at Leiden a list of offensive assertions attributed to Coccejus and Wittichius.[21] Thus confronted, the curators delegated the theologians Frederick Spanheim and Antonius Hulsius, both hostile to Cartesianism,[22] to determine which of the submitted statements were the most dangerous. The result was a list of twenty-one propositions which the curators, with the personal approval of the Prince of Orange, prohibited from being taught, disputed, or dealt with within the university.[23]

The propositions condemned by this new prohibition, finally promulgated in January of 1676, were essentially theological in character. Churchmen in the immediately preceding years had expressed their displeasure with what they considered innovations in physics as well, such as the contention that the earth moved and might be a star or planet, that the world could not be proved finite, and that there were mountains and valleys and perhaps even men and animals on the moon.[24] But after the deletion of a similar proposition touching the moon which Spanheim and Hulsius had intended to include, only one of the twenty-one forbidden propositions dealt directly with matters pertinent to natural philosophy, that being the assertion that the world was infinitely extended, thereby precluding the possibility of many worlds. The prohibition against suggesting that the Scripture had indulged the misconceptions of the common people also clearly pertained to old arguments put forward in behalf of the movement of the earth, and the condemnation of the idea that all philosophy was independent of religion had ominous overtones for natural philosophers as well. Otherwise, however, physics was not immediately at issue, for the new prohibition was intended primarliy as an assault on Coccejanism and Cartesian metaphysics.

The response elicited from the Cartesian and Coccejanist sympathizers within the university was unexpected. The ecclesiastical circles had intended to strike primarily at Wittichius and Burchardus de Volder, but it was the nearly eighty-year-old Heidanus, the former friend of both

[21] *Ibid.*, pp. 312-3.
[22] *Nieuw Nederlandsch Biografisch Woordenboek*, Vol. VIII, 888-9; Vol. X, 955.
[23] *Bronnen*, Vol. III, pp. 317-21 and 259*.
[24] Cramer, *Abraham Heidanus*, pp. 124-5.

Descartes and Coccejus, who placed himself in the forefront of the resistance.[25] Trusting perhaps to his age, his thirty-two years of teaching at Leiden, and his reputation for moderation and orthodoxy, Heidanus assumed full responsibility for a book which he, Wittichius, and de Volder soon published criticizing the course of action the curators had taken.[26] Attributed to his authorship, the *Consideratiën over eenige Saecken onlanghs voorgevallen in de Universiteyt binnen Leyden,* "Considerations of Some Recent Affairs at the University in Leiden," went through three editions in the course of 1676.[27] The curators were depicted as having allowed themselves to be led into an ill-advised policy which they could never really have intended to enforce.[28] It was clear to all that Heidanus had not been the target of the prohibition, but compelled to defend their integrity and convince the conservative ecclesiastics of the sincerity of their intentions, the curators now found it necessary to dismiss Heidanus from the faculty of the university.

The dismissal of Heidanus in 1676 marked the climax of the struggle that had dominated the history of the university throughout the third quarter of the century. This last major counteroffensive by conservative ministers against the new philosophy, which was now being taught even within the academy at Geneva,[29] gradually waned, and, troubled perhaps by regretful afterthoughts, the curators at Leiden soon abandoned the campaign that had resulted in such unforeseen consequences. Wittichius and de Volder were left undisturbed, and, though discretion might still often remain the better part of valor, the Cartesian philosophy continued to be openly taught at Leiden throughout the remainder of the century.

By no means, however, did Cartesianism now reign unchallenged in the university, and certainly not in the realm of physics. To be sure, de Volder, at his own request, had been authorized to teach physics when first appointed to the faculty in 1670, but the subject was soon taken up by Senguerdius as well, who had been named professor of philosophy in 1675 specifically as a counterweight to the Cartesians.[30] The son of a former professor of philosophy at Utrecht and Amsterdam, Senguerdius had received his doctorate of philosophy at Leiden in 1667, and was two

[25] *Ibid.,* p. 153 and *passim.*
[26] *Ibid.,* pp. 153-4. Sassen, *Geschiedenis van de Wijsbegeerte in Nederland,* p. 162. Thijssen-Schoute, "Le cartésianisme aux Pays-Bas," p. 208.
[27] Dibon, "Notes bibliographiques," p. 269, n. 18.
[28] Cramer, *Abraham Heidanus,* pp. 153-4.
[29] Charles Borgeaud, *Histoire de l'Université de Genève: L'Académie de Calvin, 1559-1798* (Genève: Georg et Co., 1900), pp. 406-12.
[30] *Bronnen,* Vol. III, p. 245. Senguerdius was offering lectures and experimental demonstrations pertaining to physics at least by 1681; see the *Series lectionum* for that year, *Ibid.,* p. 269*.

years later granted the right to give public lessons as a lector.[31] "To dissent from others has never been a religion for me," he was later to write,[32] and he soon acquired for himself the reputation of an adherent of the traditionalist party. Nonetheless, his own natural philosophy proved to be a strange and incongruous blending of obstructive traditionalism, Cartesianism, atomism, and experimentalism.

Senguerdius' mongrelization of old and new became immediately apparent in the opening pages of his *Philosophia naturalis*, first published in 1680,[33] where he undertook to vindicate scholastic usages touching first principles. He reasserted form as a principle in physics, only to reduce it, however, to matter in motion, and while explaining that matter could, in one sense, justly be called non-being, he nonetheless held it to be a perfect substance whose nature derived from extension alone.[34] In such passages, Senguerdius might also appear to be seeking a reconciliation of peripatetic and Cartesian thought, but a more uncompromising traditionalism was elsewhere revealed, as in his rejection of Copernicanism as contrary to both divine authority and such astronomical evidence as the absence of stellar parallax.[35] He settled in the end, as Kyper had done years before, for the system of Tycho Brahe.

In considering the alternative celestial systems, Senguerdius had given particular attention to the refutation of the Cartesian solar vortex.[36] Since the solar vortex also entailed the movement of the earth about the sun, the vortex became for Senguerdius the key issue in the Cartesian system, and the importance he attached to its refutation is apparent in the length and detail of his argument and the abundant footnotes to Descartes himself. Though accepting the Cartesian imagery of celestial particles, Senguerdius rejected Descartes' explication of the solar vortex as internally inconsistent and incompatible with the actual behavior of bodies in swirling liquids. Within a few years, Newton was also to make the solar vortex a focus of his own assault on Cartesian natural philosophy, but in stark contrast to his great contemporary at Cambridge, Senguerdius was primarily seeking to preserve the immobility of the earth.[37]

[31] *Ibid.*, pp. 306* and 228.
[32] *Philosophia naturalis, quatuor partibus primarias corporum species, affectiones, differentias, productiones, mutationes, et interitus, exhibens* (2nd ed.; Lugduni Batavorum: Apud Danielem à Gaesbeeck, 1685), "Ad lectorem."
[33] All references will be to the second edition, cited immediately above.
[34] *Philosophia naturalis*, p. 7 ff. and 20 ff.
[35] *Ibid.*, p. 160 ff.
[36] *Ibid.*, p. 174 ff.
[37] When de Volder was presiding in the lecture hall, however, a moving earth – or Descartes' non-moving earth borne about by celestial currents – was also still

Senguerdius' criticisms of the Cartesian solar vortex touched, nevertheless, upon real difficulties that this most extreme and most elaborate version of the mechanical philosophy faced in attempting to provide an explanation of all natural phenomena. Similar difficulties were also revealed in Descartes' explanations of terrestrial gravity, a context in which the serious implications of Senguerdius' observations were not obscured by their being marshalled in behalf of a dubious cause.[38] Senguerdius there again exposed the intricate operations of Descartes' hard-working *deus ex machina,* the celestial fluid, as being inconsistent and unworkable, and precisely because Senguerdius' arguments were not particularly profound, they revealed how near at hand the internal contradictions within the Cartesian system really lay.

Nonetheless, Senguerdius himself assumed in his own physics that the phenomena of nature were to be understood in terms of particles of extension in motion, testimony to the pervasive influence of Cartesian thought. When Senguerdius offered his own explanation of terrestrial gravity, however, he drew upon more than the meager resources of extension in motion. Betraying now the influence of the atomists, he asserted that the true cause of gravity was an *impetus* to gravity found within matter itself, an impetus similar to that responsible for motion in general.[39] "Indeed, as all bodies are moved by an impressed, inhering impetus, so, likewise, are they to be considered as being moved downwards by the same." This impetus had been originally impressed in matter by God, the first cause of all motion, and was preserved in all matter equally. Reflecting as well his traditionalist understanding of motion, Senguerdius further explained that the motion of gravity was oriented perpendicularly to the earth in the same manner as all motion was oriented towards a *terminus.* Less venturesome than Kyper years before, to say nothing of Newton, he specifically excluded the consideration of supralunary gravity on the grounds that we were unable to investigate what it was, where it was, or whether there was such a thing at all.

---

openly and firmly defended at Leiden. See in particular the disputation of Gysbertus Henricus Casembroot, *Disputatio philosophica quae est de mundi systemate,* sub praesidio D. Burcheri de Volder (Lugduni Batavorum: Apud Abrahamum Elzevier, 1694). Also Samuel Koleseri's *Disputatio philosophica inauguralis de systemate mundi,* pro gradu doctoratus et liberalium artium magisterio (Lugduni Batavorum: Apud Viduam et Haeredes Johannis Elsevirii, 1681); "that system of the world delineated by the most noble Descartes," reads one of the *Annexa philosophica* at the end, "is not to be held as a mere hypothesis, but as the truth of the matter itself."

[38] *Philosophia naturalis,* p. 76 ff.
[39] *Ibid.,* p. 80.

In resorting to an internal source of motion that first recognized and then sought a distant goal, Senguerdius had posited something that was lacking in the barren image of Cartesian extension. His colleague de Volder stressed that all we perceived in matter was quantity [40] and pointed out that, since all bodies could be conceived devoid of any motion, no body could be the source of its own motion: "If indeed there were any body which was the cause of its own motion, would not the motion of this body follow necessarily from its nature like cause and effect? As a consequence, that body would be in perpetual motion and could not be conceived without it." [41] The year after Senguerdius' *Philosophia naturalis* had first appeared, one of de Volder's students applied these arguments directly to the question of gravity: if the downward motion of gravity derived from the nature of body, "extension without motion could be perceived by no one, which completely contradicts not only experience but reason as well." [42] To resort to an "impressed impetus" to explain an internal principle of motion, moreover, was "to reveal the obscure through the obscure." [43] Nonetheless, Senguerdius had argued to no little effect that the Cartesians themselves as yet lacked a satisfactory explanation of how pure extension in motion could account for the fundamental phenomenon of gravity.[44]

Senguerdius' conservatism became apparent again in his denial of the infinitude of the world, the last step, indeed, in the final disintegration of the cosmos. Descartes himself had banished the word "infinite" from the realm of physics as appropriate to God alone, and he had spoken of the world, for which he could not, in fact, conceive any limits, as being, rather, "indefinite" in its vast extent. De Raey had declined to broach the

[40] *Disputationes philosophicae sive cogitationes rationales de rerum naturalium principiis* (Medioburgi: Typis Remigii Schreverii, 1681), p. 54.

[41] *Quaestiones academicae de aëris gravitate* (Medioburgi: Typis Viduae Remigii Schreverii, 1681), p. 39.

[42] Henricus van Bronchorst, *Disputatio philosophica de vera gravitatis causa,* sub praesidio Dni. Burcheri de Volder (Lugduni Batavorum: Apud Abrahamum Elzevier, 1685), p. VI.

[43] *Ibid.,* p. V.

[44] The Cartesians, of course, were hardly willing to concede this. De Volder's student van Bronchorst also claimed in disputation that no objections could be raised against Descartes's attribution of terrestrial gravity to the secondary vortex about the earth itself and the pressures resulting within the celestial fluid; this promised the clearest and easiest explanation, van Bronchorst declared (*Ibid.,* p. VIII). This was, however, only one of at least three possible explanations that Descartes had considered. Compare articles XIX-XXIII, "Pars quarta," of the Latin *Principia philosophiae* in the *Opera philosophica* (fourth edition) of Descartes published in Amsterdam in 1664 with the equivalent passages in the French edition of 1681, reproduced in *Oeuvres de Descartes,* ed. Victor Cousin (Paris: F. G. Levrault, 1824), Vol. III.

subject in his *Clavis philosophiae naturalis,* but Geulincx abandoned Descartes' reserve and declared that matter, at least, was "infinitely" extended.[45] If it were considered to have an end, he argued, we would have something within but nothing without, "there would then be *intra* without *extra,* and that is impossible."

Geulincx emphasized, however, that it was matter and not the world of which he spoke, for the world comprised motion as well, and whether motion also stretched to infinity, he said, we could not know.[46] If there were limits to the spread of motion, however, and this he pointedly refused to confirm or deny, these were the limits of the world as well. Beyond these limits there would then be infinite space, but space that was "completely solid, completely dark, and harder than any adamant"; [47] here was an image of naked and undisturbed Cartesian extension.

Adhering more closely to Descartes' original approach and ignoring Geulincx's distinction between extension and the world, de Volder asserted that whether the world was infinite could not be known for certain, but it seemed, in fact, most probably the case, and he could think of no limits beyond which he could not conceive still more extension.[48]

> I shall say, rather, that the world is *indefinite,* and point out that I mean by this only that I do not know with certainty, beyond all doubt, whether the world is finite or infinite, [whereas] I do know with certainty that if there are perhaps limits, I do not perceive them.[49]

Senguerdius, nonetheless, clung persistently to the finite limits of the world just as he had clung to its stable center, the earth. He argued rather precariously in his turn that since the parts of extended matter were finite, then the whole of matter must be finite, for "neither from the multiplication of finite parts can an infinite whole result nor can an infinitude admit a finite part." [50] As reflected in the prohibition of 1676, however, what troubled conservative piety about an infinite world was that it denied God the possibility of having created other worlds, a limitation on his power considered impious by scholastics since the

---

[45] *Sämtliche Schriften,* Vol. II, p. 493.
[46] *Ibid.,* pp. 495-6.
[47] *Ibid.,* p. 496.
[48] *Exercitationes academicae, quibus Ren. Cartesii philosophia defenditur adversus Petri Danielis Huetii Episcopi Suessionensis Censuram philosophiae Cartesianae* (Amstelaedami: Apud Arnoldum van Ravestein, 1695), p. 89 ff. in the second series of the pagination.
[49] *Ibid.,* pp. 90-1.
[50] *Philosophia naturalis,* pp. 12-3.

thirteenth century.⁵¹ The traditional doctrine that Senguerdius now urged to circumvent this difficulty posited, however, a more baffling infinitude of dimensionless emptiness and underscored the philosophical difficulties that also attended the idea of empty space.

For an intellect that conceived all existing physical reality as either substance or accident inhering in substance, the concept of empty space posed a disturbing paradox. It was an impasse that confronted both Cartesians and scholastics alike, those who held substance to be extension itself as well as those who conceived substance as natural bodies. To Kyper, who had in fact granted the probability of smaller vacuums, space was still nonetheless non-being,⁵² and de Volder asked if empty space was not, in truth, extended nothing, to him a patent self-contradiction.⁵³ Defined as containing nothing, empty space could have no substance, and as long as substance remained the conceptual prerequisite to reality, empty space could be granted no reality.

De Volder was particularly explicit. Since spatial extension could be imagined and perceived – and what could be perceived more clearly, he asked ⁵⁴ – the reality of space could not, in truth, be denied; but space, in turn, could never then be devoid of substance; it could never consist of nothingness alone.

> So as to linger on this matter no longer, either some properties are granted to empty space or none at all. If the latter, what is more evident than that empty space is nothing, and so he who asserts the existence of empty space asserts the existence of nothing. And what could be more absurd? But if some properties are granted, these surely will be length, breadth and depth; indeed, we perceive nothing else in space. And since that length and breadth cannot be nothing, they will assuredly be either substance or accident. If the first, since length, breadth and depth is corporeal substance, body indeed will be present; if the second, that length and breadth will surely be in body.⁵⁵

In either case, empty space was not really empty but filled with matter. Matter had been emptied of all but dimension by the Cartesians; reciprocally, they had rendered all emptiness matter.

Senguerdius, on the other hand, staunchly affirmed the existence of empty space, but it was an empty space that could not shake free of a puzzling negative conceptualization. Space, he declared, was the capacity to receive and contain bodies, though it was irrelevant whether those

---

⁵¹ Edward Grant, "Medieval and Seventeenth-Century Conceptions of an Infinite Void Space beyond the Cosmos," *Isis*, Vol. LX (1969), pp. 48-51.
⁵² *Institutiones physicae*, Vol. I, pp. 59 and 306 ff.
⁵³ *Disputationes philosophicae de rerum naturalium principiis*, p. 141.
⁵⁴ *Ibid.*, p. 145.
⁵⁵ *Ibid.*, pp. 119-20.

bodies were present or not.⁵⁶ Underlying all things but indifferent to whether or by what it was occupied, this was a space that existed before, with, and after all bodies created by God. Nonetheless, declared Senguerdius, this space had no reality of its own; though it existed, it was unreal.

When Senguerdius turned to the question of the possible creation of other worlds, he stretched this unreal but existing space far beyond the finite limits he had imposed on this world. We were not to trouble ourselves, he wrote, about the empty space in which new worlds might be produced, for any arguments against such space could be turned as well against the space in which our own world was once created, and since the very existence of our world demonstrated that the space preceding it had entailed no contradiction or absurdity, it would be no more contradictory or absurd for new bodies to be created where now there were none.⁵⁷

This unreal space in which other worlds could be created, however, apparently suggested no dimensions or extension to Senguerdius, as it had not to his medieval predecessors.⁵⁸ He felt no need to reconcile this further space with the finitude of the world or with his continuing identification of extension with the essence of matter. Nor did this empty space have any counterpart in his imagery of the real physical world; he still asserted the "circular" motion required by the Cartesian plenum, wherein every motion necessarily entailed a closed chain of compensatory motion,⁵⁹ and assumed the presence of the Cartesian subtle matter in the vacuums created in pneumatic experiments.⁶⁰ Years before, Burgersdijck's predecessor, Jacchaeus, had placed beyond the cosmos a nightmare space that was not truly "empty" because it lacked the capacity to contain anything,⁶¹ and Geulincx had described a curiously similar space of impenetrable extension. Senguerdius had now stretched beyond the world an unreal, dimensionless space of pure capacity.

Related to the philosophical problem of space was that of the relativity of motion, and here also the exchange between Senguerdius and the Cartesians at Leiden reflected the paradoxes that the new science seemed to force upon philosophy. Continuing where de Raey had left off in his

⁵⁶ *Philosophia naturalis*, pp. 152-3.
⁵⁷ *Ibid.*, pp. 156-7.
⁵⁸ Grant, "Medieval and Seventeenth-Century Infinite Void Space," *passim*.
⁵⁹ *Philosophia naturalis*, p. 49 ff.
⁶⁰ *Inquisitiones experimentales* (2nd ed.; Lugduni Batavorum: Apud Cornelium Boutesteyn, 1699), p. 15; *Rationis atque experientiae connubium* (Roterodami: Apud Bernardum Bos, 1715), *passim*.
⁶¹ *Institutiones physicae* (rev. ed.; Lugduni Batavorum: Excudebat Vidua Ioannis Patij, 1624), p. 92.

Clavis philosophiae naturalis, both Geulincx and de Volder taught their students this doctrine of motion that had at times escaped the grasp of Descartes himself and subsequently repelled even Newton. Geulincx defined motion now as the joining of nearness and remoteness between the same two objects, which were now near, that is, and now remote;[62] "for example, if A is near B and the same A becomes distant from the same B, this itself will be motion, and A is understood to withdraw from B and B from A." This motion was mutual; it was to be attributed to both A and B alike, and, hence, when a stone fell from the top of a tower, the motion was no less in the top of the tower than it was in the stone.[63]

In further arguing the mutuality of such motion, Geulincx revealed the relationship that existed at least in his own mind between the relativity of motion and the Cartesian conception of space as substance.[64] When the stone fell from the tower, those who considered only the stone to move did so, he stated, because they fancied space to be something immobile and fixed; that which they imagined to be migrating from one part of this space to another they held to be in motion, while that which remained steadfast in the same part of space they declared to be at rest.[65] It was this venerable preconception that space was immobile in all its parts and was penetrated now in this part and now in another, he declared, which obscured the true mutuality of motion.[66] In truth, it was impossible for extension to be immobile with respect to its parts or for extension to penetrate extension. "Space is body," [67] body which he had elsewhere described as completely solid and harder than any adamant. In Geulincx's Cartesian world, it was space itself that moved, and the changing relationship between its parts was all, ostensibly, that remained of motion.

Senguerdius, however, had conceived of a space that was indifferent to the coming and going, the presence or absence, of bodies, and such an understanding of space also had its implications for the understanding of motion. Motion, wrote Senguerdius, was not to be defined as a passage (*translatio*) of a body from place to place but from space to space.[68] A body could never abandon the place it filled, he explained, but,

---

[62] ... *conjunctio viciniae atque distantiae ejusdem ad idem* (*Sämtliche Schriften,* Vol. II, p. 496).

[63] *Ibid.,* p. 498.

[64] The potential relationship between the relativity of motion and concepts of full or empty space was first suggested to me by Richard S. Westfall.

[65] *loc. cit.*

[66] *Ibid.,* p. 390.

[67] *Ibid.,* p. 371.

[68] *Philosophia naturalis,* pp. 34-5.

"through motion, a body abandons the space which by virtue of its presence had been place and acquires other space, which is changed by the entrance of the body from space to place." [69] A body's place, he was saying, remained inseparable from the body itself, but when the body moved, the space it had occupied was left behind. Here was a space that was not only open and indifferent to the passage of bodies but was also immobile, and, though Senguerdius was to reject the relativity of motion on other grounds, such an immobile space, as in the Newtonian universe as well, was an environment hostile to the concept of the relativity of motion.

In 1675, in disputation, a student of de Volder's had pondered some of the puzzling implications of the relativity of motion.[70] Local motion, he had declared, was a change in the relationship between bodies, but since a body might change its position with respect to some bodies while not with respect to others, it followed that one and the same body might simultaneously be moving and at rest. All that could be done was to select one particular relationship as decisive, and the most appropriate was that between a body and the bodies immediately contiguous to it. He added, however, that the decision as to which of two mutually receding bodies one chose to say was moved and which at rest was left to the discretion of the observer.

Senguerdius could not accept all this with such tolerant agreeableness. What disturbed him most was that the relativity of motion destroyed the simple unity of motion and the possibility of locating it with certainty in a single body. As an accident, motion had to have a subject in which it existed, but the relativity of motion distributed precisely the same motion among several different bodies, attributing it to many subjects but leaving it without a subject, without something, that is, in which it might inhere.[71] Motion simply could not exist simultaneously in several subjects, "for that which is one cannot, by reason of different subjects, possess a divided essence." [72] Again revealing the continuing grip of the identification of motion with a *terminus ad quem*, Senguerdius further objected that two bodies, if their motion were mutual, would be proceeding towards opposite *termini* by virtue of the very same motion. To thus deprive motion of its unity and deny it any specific thing in which to reside was, to Senguerdius, to challenge its very reality.

[69] *Ibid.*
[70] Johannes Bruno, *Disputatio physica de motu, tertia et ultima,* sub praesidio D. Burcheri de Volder (Lugduni Batavorum: Apud Viduam et Haeredes Joannis Elsevirii, 1675), pp. II-V.
[71] *Philosophia naturalis,* pp. 29-31.
[72] *Ibid.,* p. 29.

Though his illustration of the point was not sound, Senguerdius also rightly argued that the relativity of motion was incompatible with the equally Cartesian concept of a constant total quantity of motion in the world, a concept which Senguerdius himself accepted as "probable" on the basis of experience.[73] The Cartesians at Leiden never apparently recognized this inconsistency within their physics, and Geulincx asserted that not only motion but rest as well was maintained in a constant quantity.[74] A corollary in a disputation defended under de Volder in 1692 likewise asserted that God had indeed produced a certain quantity of motion, though "whether he keeps it always the same or equal is uncertain." [75] Like Senguerdius, de Volder also appears to have continued teaching a variant of the Cartesian laws of impact, in which the relativity of motion was also overlooked.[76] Clearly, it was not the traditionalist natural philosophers alone who found the relativity of motion, the new motion in its most radical form, philosophically elusive.

At Leiden, the bitter clashes over contending philosophies and their theological implications had rapidly subsided after 1676, and the general tenor of violence within the university gradually abated as the seventeenth century drew to a close.[77] Nonetheless, fundamental questions pertinent to the changing conception of nature and the universe remained very much at issue. The new philosophy had inspired a hope that the innermost secrets of nature were soon to be revealed, but the most basic features of the universe continued to elude a clear and complete understanding. The universe was well on its way to infinity, but infinity was a concept which was generally recognized as beyond conception. Efforts to render space itself a substance which could account for the phenomena of nature still fell short of their goal, while empty space as a physical reality resisted comprehension. Motion was losing its sure philosophic

[73] *Ibid.*, pp. 37 and 43 ff. Senguerdius argued that since the quantity of motion lost by one body which moved another would be acquired not only by the body moved but by its neighboring bodies as well, twice as much motion would result.

[74] *Sämtliche Schriften,* Vol. II, pp. 513-5.

[75] Jacobus Erckelens, *Exercitationum philosophicarum tertia et vicesima, quae est de corpore,* sub praesidio D. Burcheri de Volder (Lugduni Batavorum: Abrahamum Elzevier, 1692), corollary II.

[76] See Hermannus Schuyl, *Disputatio philosophica inauguralis de vi corporum elastica* (Lugduni Batavorum: Apud Abrahamum Elzevier, 1688); this *pro gradu* disputation is dedicated to de Volder as the "principal founder of my studies." See also Senguerdius, *Philosophia naturalis,* p. 42 ff. It has been observed that Descartes' laws of impact were incorporated into the traditional physics with comparative ease; Pierre Boutroux, "L'Enseignement de la mécanique en France au XVIIe siècle," *Isis,* Vol. IV (1921-2), pp. 286-7.

[77] D. van Arkel, "Leids Studentenleven in de 16e, 17e, en 18e Eeuw," in *Geschiedboek van het Leidsche Studenten Corps* (Leiden: H. E. Stenfert Kroese, 1950), p. 28.

footing, and the ultimate comprehensibility of physical substance itself was soon to be despaired of in Newtonian science. Even the return of greater philosophic tolerance and reasonableness could not resolve these difficulties, difficulties with which the new science would have to contend as long as it was expected to satisfy the demands of philosophy.

CHAPTER VI

THE PRACTICE OF PHILOSOPHY

The question of "method" in philosophy had become in the sixteenth century the subject of much thought and debate.[1] Influenced by the pedagogical concerns of humanist reformers, a great deal of this discussion pertained to the organization and presentation of knowledge previously acquired, but a growing disenchantment with that knowledge itself also provoked reconsiderations of its initial acquisition and validation, reconsiderations that found particularly influential expression in the early seventeenth century in the writings of Bacon and Descartes. The "method" with which they were preoccupied was quite a different enterprise from the pedagogical "method" of Burgersdijck; theirs was rather a program for acquiring new knowledge and guaranteeing its reliability. For Descartes, his method and its guarantee of certainty was the rock on which the remainder of his philosophy rested, and when his philosophy was taken up in the schools, his method inevitably became a central issue in the lecture halls as well. Though that method also proved a devastating weapon in the war between philosophies, it was prized above all for its promise of sure and unclouded understanding.

Purposefully setting himself against the rising tide of scepticism in Europe,[2] Descartes had sought that certainty which only rational demonstration from indubitable first principles could ensure. The bias of his method was towards the purely mental,[3] and he opposed the powers of the isolated mind to the uncertain experience of the senses. His assurance that demonstrable certainty was attainable in philosophy exercised a profound attraction on those who likewise refused to acquiesce in either

[1] See Gilbert, *Renaissance Concepts of Method*. Also Richard McKeon, "Philosophy and the Development of Scientific Methods," *Journal of the History of Ideas*, Vol. XXVII (1966), pp. 3-22.
[2] See Richard H. Popkin, *The History of Scepticism from Erasmus to Descartes* (Rev. ed.; Assen: Van Gorcum en Comp. N.V., 1964).
[3] This, admittedly, is easily and often overstated. For a corrective, see in particular Alan Gewirtz, "Experience and the Non-Mathematical in the Cartesian Method," *Journal of the History of Ideas*, Vol. II (1941), pp. 183-210.

the inaccessibility of truth argued by the sceptics or the baffling logical entanglements in which the late scholastics appeared to be caught. His rigorously deductive emphasis spoke as well to the desire for systematization, a desire accentuated in the schools by the appreciation of the pedagogical advantages of such system. Even within the schools, however, the claims and aspirations of Cartesian rationalism would not go unchallenged, and as the seventeenth century drew to a close, the challenge, especially at Leiden, would draw its inspiration increasingly from the achievements of the new science, which the Cartesian philosophy could no longer claim to adequately represent.

The Cartesian methodological arguments had been vigorously set forth in the opening chapters of de Raey's *Clavis philosophiae naturalis*. Sense experience, he had urged, might suffice for the needs of the multitude, but the philosopher must strive for knowledge which is more certain. Had the senses not deceived us with regard to the size and motion of the heavenly bodies and the shape – and he might well have added motion – of the earth itself? [4] What then could be expected when the minute bodies that everywhere filled nature were in question? Nor did the senses provide an understanding of even the most immediate sensations, such as color, light and taste, for even lunatics and infants would then grasp what caused the diversity of such sensations, something that still baffled many philosophers. The senses did contribute, he acknowledged, to that prior and necessary body of evident knowledge from which further knowledge was deduced through reasoning and discourse. They conveyed the qualities, magnitudes, shapes, structures, positions, motions, changes and activities of bodies for the intellect to study, and they elucidated and confirmed the truth of certain concepts. Nonetheless, the true demonstration of the axioms on which a more profound and certain study of nature depended was to be derived not from the senses, but from "the internal light of the mind alone." [5]

For de Raey, a more profound comprehension of nature had its beginnings in certain "rare" ideas, ideas generally unrecognized and even denied by those who relied too much upon their senses.[6] These ideas were usually perceived only by the trained intellect, but their actual perception required no mental exertion at all. De Raey wrote of these ideas as pertaining to the *intelligentia*, and *intelligentia*, he explained, was understanding achieved neither through the bodily senses nor through laborious reasoning, but merely through the mind's effortless

---

[4] *Clavis philosophiae naturalis*, p. 21 ff.
[5] *Ibid.*, p. 41.
[6] *Ibid.*, pp. 45-6.

attention (*facili solius mentis intuitu*).⁷ It was nothing other, in fact, than a certain sense within the mind itself, "a sense of vision, as it were, belonging to the intellect," ⁸ and it was therefore frequently referred to, he added, in terms relevant to seeing, perceiving and observing.

> Indeed, just as a bright body placed before our eyes when they are open and looking in the right direction is necessarily perceived and recognized by us, so also that which is exposed to the understanding (*intelligentia*) is understood and received without exertion when we regard it with due attention and are not blinded by preconceptions.⁹

Despite his distrust of the corporeal senses, de Raey likened this power of intuitive comprehension to a sense of vision within the mind. It was an analogy consistent with the tendency of many Cartesians to identify the "clear and distinct ideas" they sought in natural philosophy with the vivid images of their imagination.

The Cartesian emphasis on the purely mental was expressed even more strongly by de Volder. The criterion of truth, he asserted, was not to be sought in things existing outside our thoughts, but in our thoughts themselves.¹⁰ The nature and attributes of things were not to be learned from things but from our concepts of them. The only possible source of true knowledge was, for de Volder, the Cartesian "clear and distinct idea," and if such clear and distinct concepts were not a confirmation of truth, then there could be no such confirmation and we were trapped in perpetual ignorance.¹¹ To prove, however, that such concepts were indeed a guarantee of truth, de Volder turned, as was characteristic of the rationalists of the age, to mathematics:

> Either, indeed, this [clear and distinct] perception makes us certain, or I will be able to know nothing certain. But I do know with certainty that two and three are five, and to that extent, I know I can possess certain knowledge of something. I know, therefore, this perception makes me certain.¹²

So efficacious a guarantee of truth were these clear and distinct ideas that nature was bound to conform to the necessary consequences they entailed.

> The clear and distinct idea clearly represents to me the nature of that which I am

---

⁷ *Ibid.*, pp. 36 and 45 ff.
⁸ ... *nil nisi sensus quidem purae mentis et quasi intellectualis visio est* . . . . (*Ibid.*, p. 36.)
⁹ *Ibid.*
¹⁰ *Exercitationes academicae adversus Petri Danielis Huetii Censuram philosophiae Cartesianae*, p. 48.
¹¹ *Ibid.*, pp. 68-9.
¹² *Ibid.*

considering, and therefore whatever is contained in the representation of that nature and whatever necessarily follows from it is contained in the nature of the thing as well and [must] follow by the same necessity.[13]

True to both his peripatetic predecessors and the empiricism of the new science, Senguerdius rejected a program for philosophizing that so exalted the powers of reason and belittled the role of the senses. In the introductory passages of his two later books, the *Inquisitiones experimentales,* first published in 1694,[14] and the *Rationis atque experientiae connubium,* published in 1715,[15] Senguerdius argued that reason by itself was incapable of understanding nature. Without the aid of the senses, knowledge of material things was simply unattainable, and he who relied only on his intellect in the examination of natural phenomena labored in vain.[16] The efforts of the mind alone, isolated from the senses and reflecting only within itself, were inadequate to ascertain the specific nature of bodies, their diversity, operations or effects.[17] "In order to penetrate the secrets of nature so that the mind can comprehend the method and laws by which nature produces these effects, the union of sense and intellect, of experience and reason, is surely necessary." [18] The very foundations of natural philosophy, after all, had been laid by experience and observation.

Earlier, in the *Philosophia naturalis,* he had already challenged the attribution of truth to concepts just because they seemed clear and convincing to the mind. Drawing aid from the arsenal of the sceptics, Senguerdius pointed out that we do sometimes believe in the reality of images that have no real existence outside our minds, and he added that he might, indeed, perceive some concept contrary to that of the Cartesians, a conception of space that was empty and absolutely nothing.[19] "Wherefore, the measure of the truth of a conception or thing is not the conception, but the thing itself, to which the conception is to be accommodated, not the other way around." [20]

In his later years, de Volder himself was to acknowledge that reason could not alone be called upon to reveal the secrets of all the sciences and that physics in particular demanded an ultimate recourse to sense experi-

[13] *Ibid.,* p. 64.
[14] All subsequent references will be to the second edition (Lugduni Batavorum: Apud Cornelium Boutesteyn, 1699).
[15] Roterodami: Apud Bernardum Bos.
[16] *Inquisitiones experimentales,* pp. 6-7.
[17] *Rationis atque experientiae connubium,* "Ad lectorem auctor."
[18] *Ibid.*
[19] *Philosophia naturalis,* p. 154.
[20] *Ibid.*

ence.[21] It was a common defect of all disciplines concerned with bodies that, without the aid of the senses, it could not be established whether that which was determined by reason corresponded, in fact, to the actual nature of things.

In metaphysics and mathematics, which deal with ideas alone – and these clearly and distinctly perceived – certain reasoning thrives at last. In these, reason rules supreme. But in physics, no matter how certainly we may draw conclusions from an hypothesis, it remains uncertain whether the bodies we have assumed in our reasoning truly exist or not.[22]

This insufficiency of reason in physics, however, was for de Volder decidedly a "defect," [23] preventing that science from achieving the certainty and complete comprehensibility of a science of pure rationality. Indeed, he even now continued to maintain that all true knowledge, harboring no doubt or uncertainty, could only derive from clear and distinct ideas provided by the mind alone.[24] And despite his acknowledgement of the deficiency of physics as a rational science, de Volder had continued to seek such knowledge in physics throughout his academic life.

De Volder's persistent commitment to the Cartesian quest for certain and uncompromised knowledge was shared by many in the seventeenth century and doubtless reflected an anxiety symptomatic of an age of such profound intellectual upheaval. John Donne and Blaise Pascal offered eloquent testimony to the malaise occasioned by a new universe that was appearing stripped of the values and familiar meaning of the old. The conviction that this new universe could be understood and knowledge of its workings acquired was precious reassurance against nature's becoming eternally alien and unpredictable.

To inspire the needed confidence in its reliability and philosophical soundness, however, such knowledge had to satisfy deeply-ingrained traditional expectations; it had to rest ultimately upon an understanding of the first and most general foundations of physical being. Only in knowledge that reached back to the first causes or principles of things, affirmed de Volder, could the mind, which he likened to a tossing fever victim,[25] justly come to rest.[26] The conviction of Cartesians like de Volder

---

[21] *Oratio de rationis viribus, et usu in scientiis* (2nd ed.; Lugduni in Batavis: Apud Fredericum Haringium, 1698), pp. 3 and 14-5.
[22] *Ibid.*, p. 29.
[23] *Ibid.*, p. 14.
[24] *Ibid.*, p. 3 ff.
[25] *Oratio qua ... sese laboribus academicis abdicavit* (Lugduni Batavorum: Apud Cornelium Boutestein, 1705), pp. 12-3.
[26] *Disputationes philosophicae de rerum naturalium principiis*, p. 2.

that such knowledge of the natural world was in fact attainable followed from their prior conviction that the principles of Cartesian physics – extension and motion – were both the real principles of physical reality and, at the same time, totally comprehensible. Because the ideas of extension and motion were perceived by the mind, it was believed, without vagueness or obscurity, and because they also corresponded, ostensibly, to the true essentials of physical existence, they were the initial "clear and distinct ideas" from which a true, certain, and completely comprehensible science of nature could alone be built. Consequently, the demonstration of the validity of these principles was of paramount concern to the proponents of Cartesian physics at Leiden. It had been the primary purpose of de Raey's *Clavis philosophiae naturalis,* and was, likewise, of de Volder's *Disputationes philosophicae sive cogitationes rationales de rerum naturalium principiis,* published in 1681.

"In my opinion," wrote de Volder in the *Disputationes,* "there is no one of sound mind who will not agree that the primary concern of the physicist – he, that is, who seeks to uncover the causes of phenomena – rightly concerns first principles or causes." [27] He provided therewith four criteria for the determination of what were to be held as real principles. The first criterion demanded that the principles be clearly and distinctly perceived; as he further explained: "I do not require that they be demonstrated or shown to be certainly true, but only that they be so perceived that it is understood which things and what sort of things they are." [28] They must, that is, be fully understood, for, as he reiterated throughout his career, nothing certain could be deduced from what itself was uncertain, nothing clear from what was obscure.[29] His second criterion – more a definition – was that the principles could not be the effects of further natural or corporeal causes, and the third, embracing a cherished tenet of the mechanical philosophy, denied them any property of mind or thought. The fourth criterion, which he considered the most important of them all,[30] required that all phenomena of the world be able to be deduced from any true principles.

The first and last, indeed, appear to have been the decisive criteria in confirming that any true and certain knowledge in natural philosophy was to be obtained only from the Cartesian principles, to which de Volder turned after having eliminated the principles of the peripatetics, the

---

[27] *Ibid.,* p. 12.
[28] *Ibid.*
[29] *Ibid.,* p. 13; *Disputatio medica, inauguralis, de natura* (Lugduni Batavorum: Apud Severinum Matthiae, 1664), § I; *Oratio de rationis viribus,* pp. 4-5.
[30] *Disputationes philosophicae de rerum naturalium principiis,* pp. 96 and 150.

"chemists," the classical atomists, and Anaxagoras. With respect to the first criterion, what was more clearly and distinctly known, asked de Volder, than extension, and who failed to comprehend a changing distance between two bodies?[31] Regarding the phenomena, it was not really necessary, he had earlier said, that all the phenomena be actually deduced – that was in fact impossible since the phenomena were infinite in number – but that it be demonstrated that all *could* be deduced. This he did to his own satisfaction by arguing, on the one hand, that the principles of matter and motion were sufficient to explain any sense perception and, on the other, that they alone could be perceived by the intellect.[32] In so satisfying his last criterion, de Volder made no appeal to the actual observation of the phenomena themselves. Having relied on reason alone, he had preserved intact a purely rational foundation for natural philosophy.

This foundation, however, was not as solid as it seemed. In contending that the Cartesian principles alone could be perceived by the intellect, de Volder had further observed that if there were any physical phenomena that did not derive from these principles, these phenomena exceeded the powers of our understanding, a possibility the seriousness of which he apparently did not as yet fully recognize.[33] Moreover, de Volder had also tentatively put forward and then withdrawn a fifth and last criterion, the retraction of which presaged the inevitable frustration of his efforts. The rejected criterion would have required that the principles also be demonstrated to be true, but, on reconsideration, de Volder doubted whether this was really necessary – or possible, it seems.[34] For what, he pondered, if the universe and all its phenomena could have come about from any one of many different causes? Then, indeed, the determination of certain truth was beyond the capability of the mind, and the most that could be done was to bring one or several of these different possibilities to light. Descartes himself, with uncertain sincerity, had declined to claim necessary truth for the elaborations of his system, but de Volder was now speaking of the very foundations of that system, those first principles from which any clear knowledge attainable in physics had to be derived.[35] Was it not an admission of the ultimate futility of pursuing what he had designated "the primary concern of the physicist," for why seek principles for certain demonstration if the physical truth of the principles themselves was uncertain? It was a

[31] *Ibid.*, p. 145 ff.
[32] *Ibid.*, pp. 17 ff and 150 ff.
[33] *Ibid.*, p. 168.
[34] *Ibid.*, p.18 ff.
[35] *Ibid.*, p. 13.

difficulty which did not as yet strike de Volder as that consequential, for he maintained that no one denied that the Cartesian principles really existed in nature.[36] But this was only to evade the problem, a problem that further revealed the doubtful prospects of the quest for the certainty of pure rationality in physics.

De Volder's intense preoccupation with the virtues of uncompromised rationality did not prevent him, however, from simultaneously pursuing other approaches to the philosophy of nature. Despite his regrets concerning the unfortunate philosophical implications of recourse to the senses, he had reaffirmed in the *Disputationes philosophicae de rerum naturalium principiis* that experience (*experientia*) as well as reason was, after all, another source of knowledge about nature.[37] And in the very same year, 1681, in which the *Disputationes* had been published, a smaller volume of de Volder's lectures, the *Quaestiones academicae de aëris gravitate,* also appeared in which he exclaimed that an elevated and beautiful thing was "philosophy founded on reason and carefully obtained experiments (*experimenta*)."[38] Though so deeply committed to the cause of Cartesian rationalism, it was de Volder who also introduced at Leiden the practice of experimental physics.

Having recently returned in 1674 from a visit to England, de Volder had petitioned the curators that he be allowed to institute a course in experimental physics at the university.[39] Funds had been provided the following year for the acquisition of instruments, and a house was purchased and converted into an auditorium for experimental demonstrations. Senguerdius, who now joined the faculty, took up the teaching of experimental physics as well, and together, de Volder performing experiments on Mondays and Senguerdius on Tuesdays, they established

---

[36] *Ibid.*, p. 168.

[37] *Ibid.*, pp. 19-20.

[38] *Quaestiones academicae de aëris gravitate,* "Lectori Philosopho. S." As was the case, apparently, with nearly all of de Volder's major works, the *Quaestiones academicae* was published without de Volder's knowledge or consent. A reviewer in Pierre Bayle's *Nouvelles de la république des lettres* (January 1685; 2nd edition, 1686; p. 227) observed, however, that de Volder should not feel badly about this, for there were many who would like to be so misused, including those who went vainly from one publisher's door to another seeking someone to print their works.

[39] *Bronnen,* Vol. III, pp. 298 and 301. Jean le Clerc, *Bibliotheque choisie, pour servir de suite a la Bibliotheque universelle,* t. XVIII (1709), pp. 362-64. Suringar, "De Leidsche Hoogleeraren," pp. 19-20.

At the very same time as de Volder, Carolus Dematius, Professor of Chemistry at the university, also requested that he be permitted to provide demonstrations in experimental physics in the chemical laboratory, which, he pointed out, was already in existence; like de Volder's, Dematius' request was also granted in January 1675. (*Bronnen, loc. cit.*).

experimental physics as a distinctive characteristic of the philosophic instruction offered at Leiden.[40]

In his initial statement to the curators, de Volder had apparently referred to the example offered "by other foreign academies and *illustre scholen,*" the latter generally meaning, to the Dutch, schools of higher learning that did not have the power of granting degrees.[41] De Volder's friend Jean le Clerc believed that it had been the Royal Society of England that had inspired de Volder to initiate experimental physics at Leiden,[42] and in any event, whatever the foreign examples alluded to, the introduction of the experimental procedures of the new science into the curriculum still constituted a pioneering innovation within the universities of the time. Experimental physics had found its way into the Italian universities throughout the century, but even there, no regular university laboratory was established until that at Bologna in 1690.[43] In Germany, though a course in experimental physics may indeed have been offered at Würzburg as early as 1660, the progressive and enterprising University of Altdorf, offspring of the city of Nürnberg, provided such a course only in 1683,[44] while Leiden's sister universities at Groningen and Franeker in the United Provinces took such initiatives in 1694 and 1701, respectively.[45] At Oxford, the first lectures on experimental physics began at Hart Hall in 1700, and at Cambridge, Newton's own university, seven years later in the Trinity College observatory.[46] Experiments were also demonstrated in Parisian colleges during the first decade of the eighteenth century, but the first chair in experimental physics at Paris was that established by the king in the College of Navarre in 1753.[47] Within the University of Salamanca, where Aristotle continued to reign throughout most of the century, the introduction of experimental physics was postponed until the reforms initiated in 1771.[48]

Meanwhile, at Leiden, the experimental physics introduced by de

[40] Senguerdius and de Volder, as well as Dematius, were both teaching experimental physics at least by 1681; see the *Series lectionum* for that year in *Bronnen*, Vol. III, p. 269*. See also the *Series* for 1694 and following; *Ibid.*, Vol. IV, *passim*.
[41] *Ibid.*, Vol. III, p. 298.
[42] *Bibliotheque choisie*, t. XVIII, p. 362.
[43] Ornstein, *The Rôle of the Scientific Societies*, pp. 218-9.
[44] *Ibid.*, p. 230.
[45] Sassen, *Geschiedenis van de Wijsbegeerte in Nederland,* pp. 166 and 174-5.
[46] Hans, *New Trends in Education,* pp. 47-50 and 137.
[47] Pierre Brunet, *Les Physiciens hollandais et la méthode expérimentale en France au XVIIIe siècle* (Paris: Albert Blanchard, 1926), pp. 101-2, 131 and 140. Jean Torlais, "Le Collège royal," in *Enseignement et diffusion des sciences en France au XVIIIe siècle,* ed. René Taton (Paris: Hermann, c. 1964), p. 273. Jourdain, *Histoire de l'Université de Paris,* p. 385.
[48] George M. Addy, *The Enlightenment in the University of Salamanca* (Durham, N. C.: Duke University Press, 1966), pp. 105, 124 and 126-9.

Volder and Senguerdius had made such an impact on the university community that, by the second decade of the eighteenth century, the apparatus of their experiments with vacuums appeared as the very symbol of philosophy itself.[49] The epochal series of seventeenth-century experiments dealing with pneumatics and the vacuum was, indeed, the primary, though by no means the exclusive, source of their experimental demonstrations. De Volder's *Quaestiones academicae de aëris gravitate* dealt primarily with the Magdeburg hemispheres, demonstrated before the Imperial Diet by Otto von Guericke in 1654, and Senguerdius' *Inquisitiones experimentales* and *Rationis atque experientiae connubium*, both more purely experimental in character than de Volder's *Quaestiones*, revolved as well around experiments with the vacuum pump and questions concerning the air. Both professors had air pumps constructed for them, and a new design by Senguerdius was well-known and widely used.[50]

The question of the vacuum was one that had long suggested potential "experiments" – seldom tried, apparently, but often cited – to contending philosophers.[51] Kyper had appealed to experiments to support his arguments on the vacuum,[52] and even Heereboord had referred to a few superficial "common experiences (*experientiae*)" ostensibly relevant to the subject.[53] By the later decades of the century, however, the brilliant ingenuity of the succession of experiments that had begun with Torricelli's invention of the mercury barometer now exercised a fascination which not even de Raey had been able to resist.

In the *Clavis philosophiae naturalis*, de Raey, like Descartes, had spoken of certain "rare" sense experiences (*experimenta*), among which he included experiences obtained through effort, skill and expense.[54] Citing Bacon for support, de Raey declared such experiences to be in fact of even less value to natural philosophy than common and obvious experiences. In contrast to the rare ideas which he considered so critical to

---

[49] In this decade, the title pages of student disputations became quite elaborate, and the Magdeburg hemispheres and other instruments from the university's collection regularly appeared, alongside symbols and allegorical figures of theology, law and medicine, as the emblem of philosophy.

[50] Claude August Crommelin, *Instrumentmakerskunst en proefondervindelijke Natuurkunde* (Leiden: Eduard Ijdo, 1925), p. 15; "Physics and the Art of Instrument Making at Leyden in the 17th and 18th Centuries," in *Lectures on Physics and Physiology Delivered in the University of Leyden During the Second Netherlands Week for American Students, July 5-10, 1926* (Leyden: A. W. Sijthoff, 1926), p. 32.

[51] See Schmitt, "Experimental Evidence for and against a Void," pp. 352-66.

[52] *Institutiones physicae*, Vol. I, p. 314 ff.

[53] *Philosophia naturalis cum novis commentariis explicata*, pp. 59-60.

natural philosophy, rare experiences were, for the most part, useless.[55]

Nonetheless, despite his low regard for the philosophic value of experiences obtained through art and effort, he had in his introductory letter of dedication acknowledged experiments (*experimenta*) as having been essential in shaping the recent development of philosophy, and he included within his text an enthusiastic and detailed account of various experiments with the Torricellian barometer.[56] These experiments, designed in large part to demonstrate the existence of the vacuum, had become a source of heated debate among the "modern" philosophers themselves, to say nothing of the Aristotelians. In opposition to the group of experimentalists whose leading spokesmen were Pascal and Roberval, the Cartesians insisted that a vacuum was not in fact created by these experiments; rather, a ubiquitous subtle matter seeped through the pores of the glass or metal vessels and filled what only seemed to be an empty space. Indeed, it was as a final proof of the presence of this subtle matter among terrestrial bodies that de Raey called these experiments to the attention of his readers.

In the four pages de Raey devoted to these experiments, however, the quarrel over subtle matter and the vacuum does not appear to have interested him. No arguments were addressed to the void itself, and the presence of the subtle matter was simply assumed as if no other alternative had been suggested. On the other hand, much was made of the equilibrium the experiments effected between the column of mercury and the atmosphere above, a fact that was pertinent neither to the debate between the Cartesians and the experimentalists nor to the concerns of de Raey's surrounding text. His treatment of the barometric phenomena strikes a peculiar note in the *Clavis philosophiae naturalis*, for while such an appeal to empirical investigation remains otherwise nearly nonexistent, the detailed and imaginative account of these experiments far exceeds the needs for which they had been ostensibly cited.

De Raey had further written of rare experiences that, though of little use to philosophy, they were nonetheless a source of considerable pleasure, for we were much affected by that which was new and unusual.[57] It would appear that such pleasure had played no little part in de Raey's own response to Torricelli's barometer. The professor of philosophy found the phenomenon "such a wonderful effect" that only with a note of reluctance did he finally acknowledge that his purpose at hand

---

[54] *Clavis philosophiae naturalis*, p. 43 ff.
[55] *Ibid.*, p. 46.
[56] "Epistola dedicatoria" and p. 191 ff.
[57] *Ibid.*, p. 44.

prevented his further consideration of "this secret of nature which astonishes so many nations." [58] Nature, under duress, was performing bizarre acrobatics and revealing hitherto unknown secrets for the pleasure as well as the enlightenment of the spectator. De Raey's digression from his self-appointed purposes in the *Clavis philosophiae naturalis* bespoke the seductive appeal of such intriguing new diversions.

Among the most notable of the new instruments devised by the seventeenth century for what Bacon had called the "vexation" of nature was indeed the air pump, and in his *Rationis atque experientiae connubium*, Senguerdius briefly outlined its history – ending, appropriately, with the pump he had himself designed – and in such a manner as to suggest again the fascination of the new way of carrying on natural philosophy.[59] After the invention of the barometric tube, began Senguerdius, new variations were contrived with expanded globes constructed in their upper ends, so that in these globes, emptied of the fallen mercury, could be observed those things that happened to bodies in airless space. "The observation of these astonishing things, never seen before, was a spur to those curious of the secrets of nature." [60] It raised the hope that, through this procedure, the sciences could be greatly advanced, and from thence, according to Senguerdius, followed the idea of constructing a machine which could empty the air from receptacles, so that the phenomena that occurred in a vacuum could be more clearly and accurately observed.

The advancement of the sciences was the ultimate goal, but the excitement engendered by such experimentation was difficult to distinguish from the fascination with the extraordinary, with nature performing as it was not natural for her to perform. For scientists of the caliber of Galileo, Pascal, Boyle and Newton, the experiment was a precision tool in a rigorous scientific method, but for many among the growing ranks of the enthusiasts of the new science, experimentation was a hazy ground where science and entertainment happily merged. Hence, the scientific dilettante so characteristic of seventeenth and eighteenth-century science.

Senguerdius echoed Bacon in referring to experiments as so many "artificial vexations" of nature,[61] but he also understood experiments to be nature's counterfeits as well. Experiments, he wrote as early as the *Philosophia naturalis*, are like imitations of nature, exhibiting her "performance and play" (*actus et lusus*).[62] In the *Inquisitiones experimentales*, he also spoke of them as "visible imitations of nature" and con-

---

[58] ...*naturae hoc arcanum tot nominibus admirandum*.... (*Ibid.*, pp. 196-7.)
[59] *Rationis atque experientiae connubium*, p. 2 ff.
[60] *Ibid.*, p. 2.
[61] *Inquisitiones experimentales*, p. 11.
[62] *Philosophia naturalis*, "Ad lectorem."

sidered how, by mechanical effects simulating the vicissitudes of natural things, even the obscure and invisible phenomena of living bodies might be seen and placed before the mind.[63] With reference again to the organic world – he had enclosed a "breathing" lung within one of the receptacles of his air pump – he declared in the *Rationis atque experientiae connubium* that he aimed "to provide a clearer concept of this through art and mechanical experiments imitating nature." [64]

The mechanical philosopher's image of a clockwork nature had acquired a new dimension. Nature was now indeed a machine, and, as such, had not only to respond according to the inescapable dictates of mechanical necessity when probed by inquisitive philosopher mechanics, but could be mimicked and reconstructed on laboratory tables as well. "In the same manner, to be sure, in which man-made contrivances (*artefacta*) perform their operations," de Volder also wrote, "natural things perform the same, for between the one and the other there is no distinction, except perhaps of degree." [65] Experimentation and the imagery of a mechanized nature now merged in a physics which had totally abolished the scholastic barrier between the "natural" and the man-made.

The fascination of manipulating and mimicking nature, however, was hardly enough to win the acceptance of experimentation as pertinent to the serious discipline of natural philosophy cultivated within the universities. What was decisive was quite simply the prestige that experimentalism and its practitioners had acquired in respectable society at large. The achievements of a host of "heroes" of the new science were widely acclaimed in circles enjoying considerable social esteem. Le Clerc, as we have seen, believed that it was the Royal Society that had inspired de Volder to introduce experimental physics at Leiden; or perhaps it had been the example of the Honorable Robert Boyle himself.[66] A list of such widely renowned experimentalists now came quickly to mind; de Volder could write that the Aristotelian contention that air was light, not heavy, had been overthrown by the experiments of Galileo, Torricelli, Roberval, Pascal, von Guericke, Boyle and Huygens.[67] He always considered it an honor, he declared, to set forth the experiments of "men illustrious in this art." [68]

Like de Volder, Senguerdius also had a roll call of eminent names that had confirmed the value of sense experience and close observation – if

[63] *Inquisitiones experimentales*, pp. 7, 15 and 30.
[64] *Rationis atque experientiae connubium*, p. 156.
[65] *Disputationes philosophicae de rerum naturalium principiis*, p. 79.
[66] *Bibliotheque choisie*, t. XVIII, p. 362.
[67] *Quaestiones academicae de aëris gravitate*, pp. 1-2.
[68] *Ibid.*, p. 9.

not experimentation as such – in natural science: Bacon, Harvey, Schott, Boyle, Willis, Redi, von Guericke, Swammerdam, Goedaert, Leeuwenhoek "and others" (a list that inevitably calls to mind the microscope, another of the seventeenth century's remarkable new instruments for revealing the new and astonishing in nature).[69] Senguerdius spoke of the "princes" of the arts and sciences of nature; having observed how they penetrated to the secrets of nature, he had decided when called to Leiden

> ... to follow in the footsteps of these princes of science, to walk that most secure, pleasant and easy path, and to teach those things which invincible reason and experiments, tested, as far as possible, by myself as well, would render most completely known.[70]

Nonetheless, experimentation brought with it its own ethic of personal originality and continual innovation, one source of the dynamic restlessness that was perhaps the most outstanding characteristic of the new science. Senguerdius, who aspired at least to try every experiment himself, admitted that it was easier to elaborate on the experiments of others than to invent them in the first place, but he did not fail to point out that he had improvised on the older experiments and contrived new contributions of his own.[71] De Volder also testified to the ethic of originality through his defensiveness, and, in so doing, touched as well upon a problem that would greatly exercise academicians of the distant future, the tension between teaching and the pursuit of original research.

Having assured his readers in his *Quaestiones academicae de aëris gravitate* that he had given due credit to von Guericke and Boyle in his demonstrations of the Magdeburg hemispheres, de Volder anticipated being reproached for doing what had been done before.[72] He certainly did not consider the past to be his subject, he observed, but neither did he consider it his purpose to demonstrate only what was new, "for apart from the fact that no one can do it, no matter how much he excels in learning and genius, it is completely alien to the task of teaching."[73] If anyone disapproved of anatomists and chemists who, in explaining their subjects, used experimental demonstrations that were neither new nor their own discoveries, let him attempt himself, challenged de Volder, to speak often and of nothing except what he has personally and recently brought to light. Rushing, moreover, to what was most difficult and

---

[69] *Inquisitiones experimentales*, p. 10.
[70] *Rationis atque experientiae connubium*, "Ad lectorem auctor."
[71] *Ibid.*, and *Inquisitiones experimentales*, p. 11.
[72] *Quaestiones academicae de aëris gravitate*, p. 8 ff.
[73] *Ibid.*, p. 9.

recently discovered while neglecting early foundations was simply bad instruction.

Already apparent in de Volder's anticipation of criticism was not only the prodding expectation of innovation and continual discovery in experimental activity, but also the divergence in the priorities of original science and education. Nonetheless, Leiden was quick to accommodate experimental physics within the curriculum, and the increasingly evident failures of pure rationalism would only add to the growing attractiveness of this other method of philosophizing. Cartesianism had been unable to restore a philosophic consensus even among the heterogeneous friends of the new science, and little progress was being made within the framework of the Cartesian system. The multitude of inconsistencies and misrepresentations of phenomena in the elaborations of Cartesian imagery were becoming steadily more apparent, and with the advent of the Newtonian era, basic premises of Cartesian physics – the ultimate comprehensibility of physical substance and the identification of that substance with extension – became alien to the continuing progress of the new science itself. The Cartesian search for the reassurance of philosophic certainty remained unfulfilled. The rising prestige of experimental science, on the other hand, offered a new alternative: for the fugitive comfort of certainty, the adventure of continual discovery, for a philosophical guarantee of truth, the confidence deriving from control.

Experimentalism did not mean, however, the complete repudiation of reason's final authority in matters of philosophic knowledge. Typifying the empiricist tradition, even Senguerdius asserted that, in seeking truth, the collected experiments were ultimately to be brought to the scale of reason.[74] It was the self-sufficiency of reason in natural philosophy, not its decisiveness in its collaboration with the senses, that Senguerdius vigorously rejected.

Nor did experimentation itself mean the downfall of the Cartesian imagery. In his later years, de Volder taught the experimental Cartesianism elaborated in Jacques Rohault's influential *Traité de physique*, published in 1671,[75] while Senguerdius retained the Cartesian imagery

---

[74] *Rationis atque experientiae connubium,* "Ad lectorem auctor." A truly Baconian bent towards the preliminary mass collection of observations is apparent in the second edition of this work, wherein Senguerdius included his daily observations of atmospheric conditions in 1697 and 1698. They are, indeed, the oldest such observations in the Netherlands that have come down to us; see J. P. Kuenen, *Het Aandeel van Nederland in de Ontwikkeling der Natuurkunde gedurende de laatste 150 Jaren* (Bataafsch Genootschap der Proefondervindelijke Wijsbegeerte te Rotterdam, 1919), p. 113.

[75] le Clerc, *Bibliotheque choisie,* t. XVIII, p. 398. On Rohault's text, see Mouy, *Le Développement de la physique cartésienne, passim.*

as the basis of that "reason" which he applied, with doubtful success, to his experiments. In defiance of de Volder and the scientific community at large, Senguerdius rejected the weight of the air as the cause of the phenomenon of the Magdeburg hemispheres and ascribed it, rather, to their obstruction of the necessary "circular" motion of the Cartesian plenum.[76] Neither the possibility of empty space he had maintained nor the perviousness of his vessels to the subtle matter weakened Senguerdius' attachment to this explanation. Likewise, to demonstrate how experiments were to be organized according to the needs of reason, he ordered the experiments in the *Rationis atque experientiae connubium* so as to confirm the branchy shape of the particles of the air.[77] In such cases, Senguerdius' reasoning continued to rest upon basic assumptions of the Cartesian or mechanistic imagination.

In his *Quaestiones academicae de aëris gravitate*, de Volder provided a more accurate portrayal of the reasoning that had been applied most profitably in the experimentation in pneumatics during the course of the century, a reasoning relying not on imaginings of the imperceptible but on the laws of hydrostatics. Those who failed to understand that the Magdeburg hemispheres demonstrated the weight of the air, de Volder argued, did so because they were ignorant of the mechanical laws of fluid bodies.[78] Two laws constituted the premises of his reasoning: that a fluid under uniform pressure will stand at a uniform level, and that when exposed to unequal pressure, the portion of the fluid under the greatest pressure will expel the rest. Archimedes had assumed these laws to be beyond any need of demonstration, wrote de Volder, and Galileo, Stevin, Boyle and, indeed, all who wrote about hydrostatics accepted them as fundamental. Such laws, however, generalizing the behavior of nature, were not always to appear to follow so clearly and necessarily from the purely cerebral images that de Volder had elsewhere maintained were the only reliable foundation for reasoning in natural philosophy.

The "marriage," *connubium*, of reason and experience had been heavily stressed by Senguerdius, but he had not attempted to explore the precise relationship that should exist between experience and the larger rational schemes which gave natural philosophy some semblance of conceptual coherence. The absence of this attempt reflects both the naïvete of his enthusiasm for experimentation and his relative insensitivity to problems concerning the structure of philosophic knowledge.

---

[76] *Philosophia naturalis*, pp. 55-6; *Inquisitiones experimentales*, pp. 139 and 155 ff.; *Rationis atque experientiae connubium*, p. 173 ff.
[77] *Rationis atque experientiae connubium*, p. 22 ff.
[78] *Quaestiones academicae de aëris gravitate*, pp. 12-3.

De Volder, as well, declined to elaborate explicitly on how experience – or experimentation – was to be articulated within the rational framework of the science of nature. De Volder's omission, however, appears to bespeak less naïvete or indifference to the problem than the extreme dichotomy that existed in his own efforts in natural philosophy. As long as he sought for that certainty which could only be deduced from an intuitive understanding of substance, experimentation could be of doubtful assistance. While reason continued to mean deduction from purely rational first principles, to ally it with experience of any kind was a self-defeating endeavor. Deeply affected by the hope for certain philosophic knowledge, yet equally inspired by the achievements of the new science, de Volder committed himself to two divergent enterprises, both, however, competing for recognition as the proper practice of natural philosophy.

Significantly, it was with respect to medicine that de Volder briefly suggested an interplay of larger conceptual schemes and experience that approximated the operation of the new science. Medicine itself was not a "science" but an "art," and, hence, its first concern was not philosophic truth but practical efficacy. In such a context, de Volder now found "hypotheses" lacking philosophic certainty appropriate to reason's role. He declared that medicine was now in need of an hypothesis of mechanical causes from which could follow what experience had already revealed.[79] Accommodating such an hypothesis to the structure of the entire body would admittedly be difficult, and it was not to be expected that the first hypothesis that came to mind would prove satisfactory. It was to be tested against the different activities of the body and closely examined and perfected by new reasoning and experiments. Physicians were to ascertain whether that which followed from the hypothesis corresponded with reality and whether, with a few modifications, it could not be adapted to all the activities of the body. If this could not be done, the hypothesis was to be changed.

No less suggestive of the procedure of the new science was de Volder's belief that the hypothesis he urged the physicians to seek was to be not only mechanical but mathematical as well.[80] The mathematization of nature has been conceived as the very essence of the intellectual revolution effected by seventeenth-century science,[81] but in the schools the rigid scholastic distinction between the two disciplines of physics and

---

[79] *Oratio de rationis viribus*, p. 27.
[80] *Ibid.*
[81] See Alexandre Koyré, "Galileo and the Scientific Revolution of the Seventeenth Century," *The Philosophical Review*, Vol. LII (1943), pp. 333-48, and "Galileo and Plato," in *Roots of Scientific Thought*, ed. Philip P. Wiener and Aaron Noland (New York: Basic Books, c. 1957), pp. 147-75.

mathematics had persisted, and even Descartes, otherwise one of the great architects of mathematical physics, had failed to integrate mathematics into the vast pictorial rendering of the universe that had become the alternative to scholastic physics in the lecture halls. De Raey, despite much emphasis on the clarity and soundness of mathematical and geometrical ideas, resorted to mathematics in his *Clavis philosophiae naturalis* only when explicating the laws of impact, and then a rudimentary mathematics it was. De Volder, however, resolved to teach a truly mathematical as well as experimental physics to his students at Leiden.

Though not immediately apparent in the popular experiments of the time, it was mathematics that made possible the most precise collaboration between reason and experimentation. It may even have seemed to provide de Volder's physics with an appearance of cohesiveness and unity. In fact, however, as conceived by de Volder, what was ostensibly mathematics had acquired an ambiguity of its own, reflecting rather than reconciling the disparate strains of his natural philosophy. On the one hand, he identified the application of mathematics in physics with the attainment of philosophic certainty through uncompromised rationality; on the other, it rested upon the measurement and quantification of phenomena.

As a medical student at Leiden years before, de Volder had already accepted the contention that only the mathematicians had succeeded in founding a science on sure and certain beginnings, "from whence it comes that the mathematical arts have been raised to a peak of perfection for which others aim but can scarcely hope...." [82] More than thirty years later, nearing the end of his career, he observed that mathematics concerned itself with ideas alone, ideas that were clearly and distinctly perceived and were derived only from the intellect.[83] Mathematics was for de Volder as for so many of his contemporaries the science *par excellence* of pure reason and, hence, of total comprehensibility. Such an admiration for mathematics was hardly unique to the seventeenth century, and even Aristotle had looked upon geometry as the model for other sciences. But Aristotle had nonetheless sharply delineated the separate realms of physics and mathematics, and it was precisely this separation, argued de Volder, that, together with the acquiescence to authority, had prevented physics from advancing any farther than it had in previous centuries.[84]

What is this method employed by the mathematicians, de Volder asked, which some would deny to physics?

[82] *Disputatio medica*, § I.
[83] *Oratio de rationis viribus*, pp. 7, 8 and 29.

It is surely nothing but that which true reason teaches us to use in all things. They adopt principles that are certain and so known in themselves that none can deny them. From these they demonstrate those things deduced as necessary consequences. Concerning all else, they admit their ignorance.... [85]

Why, then, is this method not to be applied in physics? Is it not possible to adopt certain and evident principles in this discipline as well? If not, if the mathematical method is not to be used in natural philosophy, declared de Volder, we must then despair of all knowledge in this science.

With the acquisition of the clear and distinct ideas of extension and motion, such a mathematical method in physics became possible. Indeed, the very idea of physical substance, extension, was itself a geometrical idea:

Let anyone consider whether he comprehends a triangle, a circle, a cube, a sphere or, indeed, a straight line. If he perceives anything in these, he perceives extension. Extension is surely in all of them, and without it, they could not be conceived. In the concept of a straight line, it is the extension of a line in one dimension; in the concept of a triangle or circle, it is the extension of a surface; in the concept of a cube or sphere, it is the extension of body.[86]

The mathematical method thus conceived by de Volder was not, however, the mathematization of nature achieved by Galileo and Newton. It was rather the familiar program for establishing pure rationalism in natural philosophy. In four successive disputations in 1682, de Volder's student Paulus Derecskei of Hungary undertook to demonstrate the origin and phenomena of the heavens according to a "synthetic-mathematical method"; [87] almost no mathematics were heard. What was "mathematical" was the progressive deduction of mental images from other mental images according to what seemed to the imagination, given the assumptions of the physical properties of extension, indubitable physical necessity. All had taken place within the mind, with no appeal to quantified phenomena.

The measurability of geometric images in physics was an auspicious by-product, but it contributed little to the mental visualization through which Derecskei strove to comprehend the physical world. As the complete expression of physical substance, these images were the means by

---

[84] *Disputationes philosophicae de rerum naturalium principiis*, pp. 10-2.
[85] *Ibid.*, p. 11.
[86] *Ibid.*, pp. 145-6.
[87] *Exercitatio philosophica, coelorum, siderumque lucidorum originem, et phaenomena, methodo synthetico-mathematicâ demonstrans* [in four parts], sub praesidio D. Burcheri de Volder (Lugduni Batavorum: Apud Abrahamum Elzevier, 1682).

which the universe was to be deduced as a necessary consequence of pure physical existence itself. In the shadow of such an undertaking, how paltry, indeed, the measurement of passing phenomena must have seemed. Consequently, this "mathematical method" was largely a method not of mathematics but of unquantified spatial images alone. Despite the glowing praises of mathematics, the success of Cartesian natural philosophy in the schools had not meant a corresponding advance in the understanding and cultivation of a truly mathematical physics.

De Volder himself, however, was too acute an observer of seventeenth-century science to fail to appreciate the mathematization of phenomena that had taken place, and he undertook to introduce the students at Leiden to this mathematical method as well. How clearly he distinguished it from the "mathematical" method of pure rationality is uncertain. Confusion of the two may have conveniently obscured the dichotomy in his own approach to philosophizing about nature. The irreconcilability of pure rationalism and experimentation could be veiled by the illusion of a uniform mathematical reason ruling everywhere. Nonetheless, an illusion it was, and de Volder's instruction at Leiden provided the students with two different depictions of a mathematical physics.

In 1682, de Volder added to his responsibilities the professorship of mathematics, and he assumed his new post with an address on the union of mathematics and philosophy.[88] Having often been compelled in the past to interrupt his lectures in philosophy to send his students back to their mathematical studies, he intended now, he declared, to join mathematics and philosophy and treat them as equal parts of one and the same science. Asserting again that their separation had been responsible for the philosophic sterility of the previous centuries, de Volder pointed to the seventeenth-century advances in the crucial realm of motion as demonstration of what had been accomplished through the alliance of these two sciences. Certainly no one, he argued as he had in the past, could be ignorant of what motion was. Nonetheless, he added, the principal properties of motion, and he cited the acceleration of free fall as discovered by Galileo, were not capable of being known without geometry. What he failed to add was that the ultimate confirmation of the mathematical properties of motion rested on the measurement of the phenomena.

Overlooking the philosophical qualifications of the physics that rested on this new quantification of motion, de Volder now argued its practical applicability. His predecessors had generally recognized that natural

[88] *Bronnen*, Vol. IV, p. 13. *Oratio de conjungendis philosophicis et mathematicis disciplinis* (Lugduni Batavorum: Apud Jacobum Voorm, 1682).

philosophy was not without its usefulness to other disciplines and endeavors, but they also agreed, the Cartesians as well as the scholastics, that practical application was foreign to the purposes of physics as a science; "science" was the cultivation of knowledge for the sake of knowledge alone.[89] Nor did de Volder appear to deviate from this conception of the essential character of science or philosophy, but he did dwell upon the practical powers offered by the new mathematical physics. Without the knowledge of motion it provided, no one, declared de Volder, could determine the quantity of water issuing from fountains and aqueducts.[90] Nor could they calculate the force of arrows, cannon balls and other projectiles. From this mathematical knowledge of motion had also come that "most elegant" doctrine of the pendulum, and from this, in turn, greater accuracy in the determination of time. The recent clocks, requiring a knowledge as well of the cycloid curve, would never have been invented had not the illustrious Christiaan Huygens "joined to the other sciences in which that great man excels a recondite knowledge of the mathematical arts." [91]

Speaking five years before the publication of Newton's *Principia mathematica*, de Volder contrasted the fruitful studies of gravity undertaken by "mathematicians" with the speculations of philosophers – philosophers of the "internal principle" school of thought, that is – on the cause of gravity. Unlike those chattering physicists who rest content with some useless knowledge of an internal principle of gravity, he declared, the mathematicians have taught us how to investigate the weight of bodies and to move the greatest weights with the slightest force. Eleven years after the appearance of the *Principia*, he would again observe that if the physicists persisted in their neglect of mathematics, they would continue arguing over the true cause of gravity forever.[92] But theirs was a problem that, by contrast, in no way hindered the "mechanics," whose ingenious discoveries were being put to practical and beneficial use.

In truth, a greater knowledge of mathematics could not resolve the problem of the "true cause" of gravity, and the philosophical soundness of the mathematical physics de Volder was now urging would have proved wanting by his own standards. Regardless of the philosophical

[89] Burgersdijck, *Collegium physicum*, pp. 1-3 and 10-1; Burgersdijck divided the philosophical disciplines into the "practical" and the "theoretical," the latter including physics. Heereboord, *Philosophia naturalis cum novis commentariis explicata*, p. 6. Kyper, *Institutiones physicae*, Vol. I, p. 6. De Raey, *Clavis philosophiae naturalis*, p. 44. Senguerdius. *Philosophia naturalis*, p. 2. See also Reif, "Natural Philosophy," p. 49 ff.
[90] *Oratio de conjungendis philosophicis et mathematicis disciplinis*.
[91] *Ibid.*
[92] *Oratio de rationis viribus*, p. 23.

underpinnings of this mathematical physics, however, the new powers that it provided over natural phenomena contrasted strikingly not only with the futility of philosophizing about an "internal principle," but with the stagnation and frustration of the Cartesian search for absolute truth as well. The era when science would yield truly significant technological fruits lay still, of course, in the distant future, and there is no reason to believe that de Volder differed from his predecessors in considering the utilitarian benefits that did accrue to be but the accidental, if welcome, by-products of true science. Nonetheless, the applicability in the realm of physical reality of knowledge acquired through this mathematical physics was already emerging as a compelling confirmation of its validity as a science of nature.

Unfortunately, the precise character of the mathematical physics taught by de Volder in his dual capacity of professor of philosophy and mathematics at Leiden remains disappointingly obscure. The only example of his own use of mathematics in natural philosophy is found in the *Quaestiones academicae de aëris gravitate,* published the year before he assumed the professorship of mathematics. Having weighed a hollow sphere filled with air, then partially emptied of air, and then partially filled with water, he calculated the ratio of the weight of air to the weight of water.[93] Elsewhere, he also noted the particular applicability of mathematics to the laws of impact,[94] and another of his students, Hermannus Schuyl, reduced a variant of the Cartesian laws to algebraic formulae in his own doctoral disputation in 1688.[95] Beyond this, however, little remains as evidence of what was undertaken in de Volder's classroom.

We cannot be certain of the sophistication of the mathematics de Volder taught, and, again, the elementary character of the philosophy curriculum would have imposed narrow limitations, but there is little doubt as to his own personal competence. Jean le Clerc, who claimed to have known de Volder fairly well, wrote that de Volder had applied himself to the new techniques of differential and integral calculus and had undertaken a close reading of Newton's *Principia mathematica* soon after it was published.[96]

[93] *Quaestiones academicae de aëris gravitate,* p. 49 ff. Senguerdius also resorted to simple quantification in his *Rationis atque experientiae connubium.*
[94] *Oratio de conjungendis philosophicis et mathematicis disciplinis.*
[95] *Disputatio philosophica inauguralis de vi corporum elastica* (Lugduni Batavorum: Apud Abrahamum Elzevier).
[96] *Bibliotheque choisie,* t. XVIII, pp. 347 and 379-80.
The university acquired a copy of the first edition of the *Principia mathematica* with the purchase of the library of Isaac Vossius, who died in 1689. The collection arrived at the university library in 1690, from which time it was available to the faculty but not, because of a lawsuit between the heirs of Vossius and the curators,

I even recall having heard Mr. de Volder remark that shortly after it appeared, the late Mr. Huygens, a great mathematician but unfamiliar with the new methods [of calculus] of which I spoke, came to see him at Leiden and approached him on the subject of Newton's book. He confessed to Mr. de Volder that he found this book extremely obscure, and asked him what he thought of it. Our philosopher answered that it was not, in fact, easy to penetrate to the principles of the author's demonstrations, but he had found those he had examined to be true.[97]

Though le Clerc's friendship for de Volder may well have colored his recollection of this anecdote, among de Volder's papers were later to be found calculations of some of Newton's proofs. That Huygens specifically designated de Volder to examine and edit his own posthumous papers, left to the University of Leiden, testifies as well to the high regard in which Huygens held de Volder's scientific and mathematical abilities.[98]

These abilities and his acute appreciation of the accomplishments of the new science placed de Volder among its more devoted and effective supporters. Far more than Senguerdius, he appears to have recognized the scope and character of the scientific achievements and innovations of his century. Nonetheless, he continued to cling as well to an extreme ideal of philosophic truth and certainty that, ultimately, was not compatible with the aims and capacities of the new science. The outcome for de Volder was inevitable, and in his last years he could no longer disguise the failure of the quest for certain knowledge to which he had been so dedicated. He became disillusioned with Cartesianism in these later years, le Clerc recalled, and complained now – despite the new science – of the meagerness of man's "lights" and the little progress that had been made in the knowledge of truth.[99]

In 1705, aged sixty-two and troubled in health, de Volder retired from his duties at the university.[100] He gave his illness as the cause, but le Clerc later suggested that also among de Volder's reasons for retirement was his

to the general public. The general public gained access to Vossius' collection only in 1705. I am indebted for this information to R. Breugelmans, Assistant Keeper at the University of Leiden Library, who communicated with me by letter in March 1972 and cited as a reference H. W. Tydeman in *Mnemosyne, Mengelingen voor Wetenschappen en Fraaije Letteren*, Vol. XV (1825), pp. 259-90.

[97] *Ibid.* De Volder and Huygens were, in fact, the first in the Netherlands to have directed their attention to the *Principia mathematica;* Sassen, *Geschiedenis van de Wijsbegeerte in Nederland*, pp. 220-1.

[98] Along with Bernhardus Fullenius, professor at the University of Franeker, de Volder was designated for this task in Huygens' will.

[99] *Bibliotheque choisie*, t. XVIII, p. 398. In his discontent with Cartesianism, de Volder also became involved in an indirect and dissatisfying correspondence with Leibniz; see L. J. Russell, "The Correspondence Between Leibniz and de Volder," *Proceedings of the Aristotelian Society*, new series, Vol. XXVIII (1927-8), pp. 155-76.

[100] *Bronnen*, Vol. IV, p. 220. Suringar, "De Leidsche Hoogleeraren," p. 20.

weariness with teaching Cartesian physics and metaphysics and his unwillingness to begin building a new system.[101] He had come to recognize how little that was certain the teachings of Descartes and Rohault contained, a loss of faith that had followed from his own reflections, believed le Clerc, and the influence of the "clever English." As evidence of the disillusionment of de Volder's later years, le Clerc recalled de Volder's address in 1698 on "The Powers of Reason"; it was in this address that, while praising the mathematical methods of Newton and Huygens, de Volder had acknowledged the inadequacy and uncertainty of reason in physics.[102]

The internal strains of a radical mechanical philosophy and the relentless advance of the new science had made the foundering of Cartesian natural philosophy unavoidable, and as the seventeenth century closed, growing numbers of European intellectuals would experience de Volder's disillusionment. Nonetheless, on the threshold of the "Age of Reason," reason was far from discredited. The natural philosopher was adopting a form of reason, mathematics, that offered a dramatically expanding freedom of manipulation and far greater logical rigor and precision than either the dialectic of the scholastics or the imagery of the Cartesians. Moreover, nature's conformity to this mathematical reason was being compellingly demonstrated by experimentation. Though mathematics could not provide a deeper understanding and the why and wherefore remained unknown, reason's rule in nature had never before found such precise and sweeping confirmation.

By the age of Newton, however, the higher mathematics applied in physics had become so abstruse as to challenge even the trained mathematician – witness de Volder struggling over the proofs of the *Principia mathematica* – and it was not mathematics, therefore, but experimentation itself that became the popular hallmark of the new science in the following decades. The manipulations of mathematics had passed far beyond the ability of common sense to follow, but the spectacle of the experiment remained a dramatic and accessible demonstration of control over nature, evidence of a knowledge of nature's ways that would eventually obscure the absence of a deeper knowledge of nature's inner content.

---

[101] *Bibliotheque choisie*, t. XVIII, pp. 398-9.
[102] *Oratio de rationis viribus*, pp. 22-3 and 29.

CHAPTER VII

's GRAVESANDE AND MUSSCHENBROEK: NEWTONIANISM AT LEIDEN

Following the retirement of de Volder in 1705, the discipline of physics at Leiden was left largely in the hands of the aging Senguerdius. Demonstrating experiments weekly and lecturing two or three days a week as well, he dominated the instruction in natural philosophy for nearly the next decade and a half.[1] The theoretical side of the physics taught at Leiden during these years was consequently somewhat retrogressive, but in Senguerdius' continuing cultivation of experimental physics, Leiden remained a leader among the European universities. In 1711, to be sure, the traveling German bibliophile Zacharias von Uffenbach found the facilities for experimental physics at Leiden hardly inspiring.[2] The demonstration hall itself was untidy – "especially for Holland" – and the instruments were found lying about unclean and sometimes broken. Uffenbach was told, moreover, that the experiments were now neither so interesting nor so well attended as those once offered by de Volder. Nonetheless, at Senguerdius' death in 1724, very few of the well over one hundred individual items in the university's collection of experimental equipment were broken or defective,[3] and much of Senguerdius' instruction in experimental physics had taken place not in the university auditorium but at home, as was characteristic of the teaching at the university in the eighteenth century.[4] It was from such lessons at home that Senguerdius

---

[1] See the *Series lectionum* for these years in *Bronnen*, Vol. IV, *passim*.
[2] Zacharias Konrad von Uffenbach, *Merkwürdige Reisen durch Niedersachsen, Holland und Engelland* (Ulm, 1754), Vol. III, pp. 425-6.
[3] *Bronnen*, Vol. IV, pp. 104*-6* and 184*-6*.
[4] *Ibid.*, p. 259. See the review of Senguerdius' *Rationis atque experientiae connubium* in the *Nouvelles de la république des lettres* (January-February 1716, p. 111), now edited by his colleague Jacques Bernard. The academic senate had informed the curators of the advantages of such informal instruction at home as early as 1700, observing that some professors sometimes dispensed with their formal public lessons altogether; *Bronnen*, Vol. IV, pp. 81*-7*. See also de Vrankrijker. *Vier Eeuwen*, p. 64, and Siegenbeek, *Geschiedenis der Leidsche Hoogeschool*, Vol. I, p. 303.

had drawn the series of experiments in his *Rationis atque experientiae connubium,* published four years after von Uffenbach's visit.[5] Though doubtless lacking the acuity of his former colleague, Senguerdius appears to have conscientiously maintained the tradition of experimental physics at Leiden, a tradition which, as von Uffenbach himself observed, other schools would have done well to emulate.[6]

Nor did the traditionalism of Senguerdius' theoretical physics prevail unchallenged at Leiden during these years. Immediately upon de Volder's retirement and on his own recommendation, the curators had found a successor in Jacques Bernard.[7] Editor of the *Nouvelles de la république des lettres* and minister of the Walloon congregation in Leiden, Bernard was appointed lector in philosophy and mathematics in 1705 and professor in these fields in 1712. He was apparently expected to continue de Volder's instruction in experimental physics alongside Senguerdius, but there is no indication of his ever having done so.[8] In the surviving *Series lectionum* of the university, he is listed as formally offering public instruction only in ethics and *pneumatica,* the science of spiritual beings, though his efforts clearly ranged more broadly.[9] Little is known of his philosophy, but von Uffenbach spoke very disparagingly of his learning (as well as his personality).[10] He left no writings of his own on physics, but student disputations suggest that, after initially espousing a Cartesian viewpoint, he was later affected by anti-Cartesian and perhaps Newtonian influences.[11] Upon his death in 1718, he was succeeded by Jacobus Wittichius, who may be considered Leiden's last Cartesian.[12] Continuing to teach until his own death in 1739, Wittichius was to be overshadowed by his colleague who had assumed the instruction in mathematics and astronomy the year before his own coming, Willem Jacob 's Gravesande.

During the first two decades of the eighteenth century, however, it was

---

[5] The title page states that the contents were *"in usum domesticae institutionis,"* while in the *"Ad lectorem auctor,"* Senguerdius remarked that the experiments had been both publicly and privately performed.

[6] *Merkwürdige Reisen,* Vol. III, p. 426.

[7] *Bronnen,* Vol. IV, pp. 221 and 259. See the *Nieuw Nederlandsch Biografisch Woordenboek,* Vol. III, 101-2.

[8] *Bronnen,* Vol. IV, pp. 216 and 221.

[9] *Ibid., passim.* Petrus van Musschenbroek, *Introductio ad philosophiam naturalem* (Lugduni Batavorum: Apud Sam. et Joh. Luchtmans, 1762), p. 2.

[10] *Merkwürdige Reisen,* Vol. III, pp. 494-5.

[11] Sassen, *Geschiedenis van de Wijsbegeerte in Nederland,* p. 227. See also Joannes Augustus Bazin, *Disputatio philosophica de formis* (Lugduni Batavorum: Apud Jacobum Poereep, 1713), and Balthazar Branchu, *Dissertatio philosophica de elementis* (Lugduni Batavorum: Apud Petrum Vander Aa, 1715), both defended under Bernard's presidency.

[12] Sassen, *Geschiedenis van de Wijsbegeerte in Nederland,* pp. 227-9.

the faculty of medicine, and more specifically the fame of Hermannus Boerhaave, that was responsible for beginning Leiden's greatest period of renown as a center of learning attuned to the more progressive trends of European scientific development. Born near Leiden, where he subsequently studied under both Senguerdius and de Volder, Boerhaave had acquired a lectorship in medicine at the university in 1701, after which his gifts as a teacher won for him a steadily growing influence within the university and throughout Europe at large.[13] Having been promised the first chair on the medical faculty that fell vacant, he was named professor of medicine and botany in 1709, and five years later he assumed the responsibility for clinical instruction as well. In 1718, he was further appointed professor of chemistry and was later to publish, in 1732, one of the century's most important textbooks on chemistry, his *Elementa chemiae*.[14] By the time of his death in 1738, his influence as a clinician and teacher was unsurpassed and his name had become one of the most widely known in European medical and scientific circles.[15]

It was of no little moment to the intellectual climate at Leiden and the university's reputation abroad that, in stressing the importance in medicine of a thorough scientific background, Boerhaave had rapidly made himself known as an advocate of the attitudes and procedures increasingly associated with the name of Isaac Newton, the "prince of all philosophers," Boerhaave declared, and "wonder of our century." [16] An outspoken admirer of the British school of natural philosophy, Boerhaave prepared the way for the emergence of the University of Leiden as a leading center of Newtonian and experimental science on the continent.[17]

Among the students at Leiden during the years of Boerhaave's growing

---

[13] *Nieuw Nederlandsch Biografisch Woordenboek*, Vol. VI, 127-41. Sassen, "The Intellectual Climate in Leiden in Boerhaave's Time," *Boerhaave and His Time*, p. 5.

[14] On Boerhaave's chemical teachings and their impact, see Hélène Metzger, *Newton, Stahl, Boerhaave et la doctrine chimique* (Paris: Félix Alcan, 1930), p. 191 ff.; Frank Greenaway, "Boerhaave's Influence on Some 18th Century Chemists," *Boerhaave and His Time*, pp. 102-13; and Milton Kerker, "Herman Boerhaave and the Development of Pneumatic Chemistry," *Isis*, Vol. XLVI (1955), pp. 36-49.

[15] Lester S. King, *The Medical World of the Eighteenth Century* (Chicago: The University of Chicago Press, 1958), p. 59 ff. I. Bernard Cohen, *Franklin and Newton: An Inquiry into Speculative Newtonian Experimental Science and Franklin's Work in Electricity as an Example Thereof* (Philadelphia: The American Philosophical Society, 1956), p. 214. Castiglioni, *A History of Medicine*, p. 615.

[16] Hermannus Boerhaave, "Oratio... de comparando certo in physicis," in *Opuscula omnia, quae hactenus in lucem prodierunt* (Hagae-Comitis: Apud J. Neaulme, 1738), pp. 28-9.

[17] Peter Gay, *The Enlightenment: An Interpretation* (New York: Alfred A. Knopf, 1969), Vol. II, p. 135.

influence were both 's Gravesande and Petrus van Musschenbroek, who were to carry on in the discipline of physics at Leiden what Boerhaave had begun. 's Gravesande had enrolled in 1704 as a student in law, and despite having early developed an interest in mathematics, he appears to have ignored the courses then being offered by both Senguerdius and Bernard.[18] Nonetheless, he continued to cultivate his scientific abilities on his own, and not without considerable success, as an *Essai de perspective* written during these years testifies. Having acquired his law degree in 1707, he began to practice in The Hague, and there joined as well in the founding of the *Journal littéraire,* to which he contributed a number of articles on science and mathematics.

In 1715, 's Gravesande and a fellow editor of the *Journal littéraire* were selected as secretaries to a special embassy being sent to honor the crowning of George I of England. The ensuing trip lasted for perhaps two years, during which time he established close and lasting ties with the English scientific community.[19] He was elected a member of the Royal Society, performed experiments with the accomplished disciple of Newton, Desaguliers,[20] and came to know Newton himself. Indeed, it was Newton's high opinion of the lawyer from the Hague that ultimately led to 's Gravesande's appointment at the University of Leiden. The great English scientist expressed his high regard for 's Gravesande to one of the Dutch ambassadors to the coronation, who passed it on in turn to the curators of Leiden as a recommendation for appointment. In 1717, accordingly, 's Gravesande abandoned his legal career to accept the professorship of astronomy and mathematics at the university, where he was to remain until his death in 1742.

's Gravesande began his new affiliation with the university with lectures on "the principles of cosmography and the physical causes of the celestial motions," supplemented, on suitable winter evenings, by practical instruction in the university observatory.[21] He initiated his own lessons in experimental physics as well and, with the death of Senguerdius in 1724, assumed responsibility for the experimental demonstrations in the uni-

---

[18] For biographical data on 's Gravesande, see the *Nieuw Nederlandsch Biografisch Woordenboek,* Vol. VI, 623-7. See also J. N. S. Allamand, "Histoire de la vie et des ouvrages de Mr. 's Gravesande," in 's Gravesande, *Oeuvres philosophiques et mathématiques,* ed. J. N. S. Allamand (Amsterdam: Marc Michel Rey, 1774), Vol. II.

[19] The embassy itself officially lasted little over a year, but 's Gravesande apparently remained a year longer; Allamand, "Histoire," p. XXI, n. L, and G. C. Gerrits, *Grote Nederlanders bij de Opbouw der Natuurwetenschappen* (Leiden: E. J. Brill, 1948), p. 164.

[20] Cohen, *Franklin and Newton,* p. 235.

[21] See the *Series lectionum, Bronnen,* Vol. IV, *passim.*

versity auditorium. In 1730, he was further charged with the instruction, in Dutch, in civil and military engineering, a course first introduced at the instigation of Prince Maurice in 1600.[22] This post was soon passed on to another, however, in 1734, when 's Gravesande and Wittichius both received the broader title of "professor of the whole of philosophy," after which 's Gravesande also began giving private lessons in logic, metaphysics and ethics.[23] In 1736, he was to publish an *Introductio ad philosophiam, metaphysicam et logicam continens,* and according to his friend and colleague (and former student), J. N. S. Allamand, was also planning, when he died in 1742, to publish a course on ethics, since he had found no books on the subject sufficiently "methodical." [24]

's Gravesande's fame, however, which prompted both Peter I and Frederick II to seek his membership in their respective academies in St. Petersburg and Berlin,[25] rested on his reputation as a physicist, and his most influential publication was his *Physices elementa mathematica, experimentis confirmata, sive introductio ad philosophiam Newtonianam,* which first appeared, in two volumes with fifty-eight large and very handsome plates, in 1720 and 1721.[26] The first edition was directed towards his students and contained a detailed description of experiments they had seen him perform in his private lessons.[27] For the further convenience of students, he published as well a portable, one-volume abridgement of his course material in 1723, under the title of *Philosophiae Newtonianae institutiones, in usus academicos.*[28] The enthusiasm with which the *Physices elementa mathematica* was received by accomplished scholars throughout Europe, however, induced 's Gravesande to bring out a second edition of this larger work in 1725, now directed more towards his own peers in the scientific world; a third edition, also considerably reworked and augmented, would appear in 1742, the year of his death.[29]

---

[22] See above, p. 10, n. 46. 's Gravesande succeeded to the chair following the death of Henricus Coets, who had carried out its obligation as a lector since 1701; *Bronnen,* Vol. IV, pp. 158-9, and Vol. V, p. 94.

[23] Allamand, "Histoire," p. XXVII, n. R. These lessons did not find their way into the formal listings of university offerings in the surviving *Series lectionum*; *Bronnen,* Vol. V, *passim.*

[24] Allamand, *op. cit.,* p. LII, n. AA.

[25] *Ibid.,* p. LIX, n. DD.

[26] Lugduni Batavorum: Vol. I, Apud Petrum Vander Aa et Balduinum Janssonium Vander Aa, 1720; Vol. II, Apud Petrum Vander Aa et B. et P. Janssonios Vander Aa, 1721.

[27] 's Gravesande, *Philosophiae Newtonianae institutiones, in usus academicos* (Lugduni Batavorum: Apud Petrum Vander Aa, 1723), "Ad lectorem."

[28] *Ibid.*

In England, where the Dutch Newtonians enjoyed an influence surpassing that of any other continental scientists in the early eighteenth century, the *Physices elementa mathematica* had been welcomed and widely appreciated as soon as it appeared.[30] Within a year of its publication, the first edition had appeared in two English translations, one by Desaguliers, which was to go through six editions before mid-century, and another, which actually displeased 's Gravesande, ostensibly under the supervision of John Keill.[31] Newton's Dutch admirers also exercised a considerable influence in France as well, despite the pervasive commitment to Cartesianism that had now replaced in France the official hostility of the preceding century.[32] The first edition of the *Physices elementa mathematica* had been ignored by the prestigious *Journal des sçavans* and was reviewed unsympathetically elsewhere, but two separate translations of the third edition were to appear in 1746 and 1747. By that time, 's Gravesande's reputation in France was already such that Voltaire, making the best of a necessary retreat from France, had sought out 's Gravesande a decade before to attend his experimental demonstrations and to solicit his opinion of Voltaire's own manuscript, the *Élémens de la philosophie de Newton*.[33] 's Gravesande also made inroads into Germany, otherwise largely dominated by the anti-Newtonian sentiments of the followers of Leibniz and Christian Wolff, and in St. Petersburg, though the Wolffian influence was quickly to prove preponderant there as well, the *Philosophiae Newtonianae institutiones* was selected as the basis for some of the earliest public lectures at Peter the Great's newly-founded academy.[34]

In the Dutch Republic itself, 's Gravesande's influence secured the growing prestige of Newtonian and experimental science throughout the provinces,[35] and the most eminent and effective of 's Gravesande's fol-

---

[29] Allamand, "Histoire," pp. XXIX-XXXII, n. S.

[30] Robert E. Schofield, *Mechanism and Materialism, British Natural Philosophy in An Age of Reason* (Princeton, N. J.: Princeton University Press, 1970), pp. 134-42.

[31] *Ibid.*, pp. 140-1. 's Gravesande, *Physices elementa mathematica*, Vol. II, "Lectori S." E. W. Strong, "Newtonian Explications of Natural Philosophy," *Journal of the History of Ideas*, Vol. XVIII (1957), p. 54.

[32] See Brunet, *Les Physiciens hollandais*, Ch. II.

[33] *Ibid.*, p. 127. Allamand, "Histoire," pp. XXXIV-XXXVI, n. T.; XXXIII, n. S.; and LIV-LV, n. BB. Ira O. Wade, *The Intellectual Development of Voltaire* (Princeton, N. J.: Princeton University Press, 1969), pp. 364-6.

[34] Allamand, *op. cit.*, p. XXXIV, n. T. Valentine Boss, *Newton and Russia: The Early Influence, 1698-1796* (Cambridge, Mass.: Harvard University Press, 1972), pp. 90 and 102-4. See also Ronald S. Calinger, "The Newtonian-Wolffian Confrontation in the St. Petersburg Academy of Sciences (1725-1746)," *Cahiers d'histoire mondiale*, XI (1968-9), pp. 417-435.

[35] Sassen, *Geschiedenis van de Wijsbegeerte in Nederland*, pp. 224-5 and 229.

lowers, Musschenbroek, was ultimately to teach at Leiden with no less acclaim and distinction than 's Gravesande himself. A member of a well-known family of instrument makers at Leiden, Musschenbroek had studied at the university under Boerhaave and graduated with a degree in medicine in 1715.[36] He journeyed to England in 1717 and, like 's Gravesande shortly before, met Newton and other members of the English scientific world. Returning to Leiden and doubtless attending the classes now being given by 's Gravesande – with whom Musschenbroek's brother, Jan, was already collaborating in building what would be one of the century's most notable collections of scientific instruments [37] – Musschenbroek acquired his doctorate in philosophy in 1719.[38] That same year he accepted a post as professor of mathematics and philosophy (and later medicine) at Duisburg in the Rhineland but four years later returned to the United Provinces as professor of mathematics and philosophy (and later astronomy) at the University of Utrecht. There, he soon won for himself an international reputation as an experimental physicist and teacher, and when Wittichius died in 1739, Musschenbroek was invited to Leiden to replace him. Assuming his professorship in mathematics and philosophy at Leiden in 1740, Musschenbroek taught alongside 's Gravesande, whom he had declared one of the greatest philosophers of the century,[39] for the remaining two years of the latter's life. With 's Gravesande's death, Musschenbroek succeeded to the responsibility for experimental physics at the university and continued as 's Gravesande's successor in this field until his own death in 1761.[40]

More prolific as an author than 's Gravesande, Musschenbroek produced a continuing series of textbooks on natural philosophy throughout his career. In 1726, while at Utrecht, he published a résumé of his lectures in an *Epitome elementorum physico-mathematicorum, conscripta in usus academicos*.[41] Three years later appeared his *Physicae experimentales, et geometricae, de magnete, tuborum capillarium vitreorumque speculorum attractione, magnitudine terrae, cohaerentia corporum fir-*

[36] For biographical data on Musschenbroek, see *Nouvelle biographie générale*, ed. J. C. F. Hoefer (Paris: Firmin Didot Frères, 1852-66), Vol. 37, 30-33. The brief sketch in the *Nieuw Nederlandsch Biografisch Woordenboek*, Vol. X, is regrettably inadequate. On the Musschenbroek family, see also Crommelin, "Physics and the Art of Instrument Making," and Maurice Daumas, *Les instruments scientifiques aux XVIIe et XVIIIe siècles* (Paris: Presses Universitaires de France, 1953) pp. 115-8 and 326-7.
[37] Crommelin, *op. cit.*, p. 35. Daumas, *op. cit.*, pp. 326-7.
[38] *Bronnen*, Vol. IV, p. 276*.
[39] *Essai de physique*, trans. Pierre Massuet (Leyden: Chez Samuel Luchtmans, 1739), Vol. I, "Preface."
[40] *Bronnen*, Vol. V, pp. 236 and 258.
[41] Sassen, *Geschiedenis van de Wijsbegeerte in Nederland*, p. 231.

morum dissertationes, containing well over two hundred experiments, most of which had been performed before his students at the university.[42] A new edition of the *Epitome* was published in 1734 and again in 1741 under the title, *Elementa physicae, conscripta in usus academicos.* Written explicitly for the use of university students, the *Elementa physicae* was translated into English in 1744 and into German in 1747.[43] In 1739, Musschenbroek had also brought out an influential Dutch elaboration of the same theme, the *Beginselen der Natuurkunde*;[44] more developed and expansive than his student texts, it was immediately translated into French and published the same year (and again in 1751) as the *Essai de physique.* 1748 brought the publication of yet another course text, the *Institutiones physicae, conscriptae in usus academicos,* similar to the earlier *Elementa physicae* in both organization and compactness. An even more abbreviated *Compendium physicae experimentalis* would appear posthumously in 1769, completed according to Musschenbroek's last writings by his long-time colleague, Johannes Lulofs.[45] The *Compendium* was derived from Musschenbroek's most important work, also completed and posthumously published by Lulofs in 1762, the *Introductio ad philosophiam naturalem.*[46] Translated into French and published at Paris in 1769 as the *Cours de physique experimentale et mathematique,* the *Introductio,* in two sizable volumes with over sixty plates, became, along with 's Gravesande's *Physices elementa mathematica,* the major work representing the eighteenth-century Dutch school of Newtonian and experimental science.[47]

In an era when the continent was still largely divided between the Cartesians on the one hand and the disciples of Leibniz and Wolff on the other, both 's Gravesande and Musschenbroek were, from the beginning of their academic careers, to be counted among the advance party of Newtonians on the continent. 's Gravesande had spoken of Newton as the "restorer of the true philosophy" in his inaugural address[48] and offered his *Physices elementa mathematica* as an "Introduction to New-

---

[42] Lugduni Batavorum: Apud Samuelem Luchtmans, 1729; "Praefatio."

[43] Editio altera; Lugduni Batavorum: Apud Samuelem Luchtmans, 1741; "Praefatio." Schofield, *Mechanism and Materialism,* p. 141.

[44] Crommelin, "Physics and the Art of Instrument Making," p. 37. Gerrits, *Grote Nederlanders,* p. 170-1.

[45] Venetiis: Apud Franciscum et Nicol. Pezzana, 1769; Lulofs' "Praefatio."

[46] Lulofs' "Praefatio."

[47] Sassen, *Geschiedenis van de Wijsbegeerte in Nederland,* p. 231. Gerrits, *Grote Nederlanders,* p. 171.

[48] "Discours sur l'utilité des mathématiques dans toutes les sciences, et particulièrement dans la physique; et sur les secours que fournit la physique pour perfectionner l'astronomie," in *Oeuvres philosophiques et mathématiques,* Vol. II, p. 321.

tonian Philosophy" three years later. Musschenbroek was less inclined to designate his physics as specifically "Newtonian," but he also, having himself lauded Newton as the "highest of mortals" in his own inaugural address at Utrecht,[49] rested his natural philosophy on the English scientist's achievements.

What precisely the "Newtonian philosophy" was remained, to be sure, somewhat uncertain.[50] His followers at Leiden often stressed his celebrated *"hypotheses non fingo,"* "I feign no hypotheses," as a decisive hallmark of his thought and, hence, of any proper Newtonian. 's Gravesande declared that, even apart from the espousal of Newton's theories and discoveries, "we justifiably call Newtonian [that] philosophy in which, hypotheses having been rejected, conclusions are deduced from phenomena; no one before Newton followed this method unremittingly or even proposed that it was always to be followed." [51] Years later, Allamand also asserted that in order to be Newtonian it was not enough merely to cultivate geometry and experimentation in physics: "It is necessary, at the same time, to reject hypotheses. This is what Descartes and his disciples have never done." [52] While Musschenbroek may not have specifically designated the repudiation of hypotheses as a criterion of Newtonianism, he also paid due respect to the clearly Newtonian injunction in proscribing all hypotheses, "conjectures," and "fictions" from physics.[53]

As conceived by the Newtonians at Leiden, hypotheses were explanations of the workings of nature that were advanced without empirical confirmation, but such hypotheses were in fact hardly lacking in the natural philosophy taught by the Dutch Newtonians themselves.[54] Even 's Gravesande, who was far more cautious than the richly imaginative Newton, still depicted the physical world in the persisting imagery of corpuscularism, one of the most typical characteristics, indeed, of eighteenth-century Newtonianism.[55] Musschenbroek, continuing likewise to portray the world as an intricate complex of tiny particles and pores,

[49] *Oratio de certa methodo philosophiae experimentalis* (Trajecti ad Rhenum: Apud Guilielmum Vande Water, 1723), p. 22.
[50] For different understandings of the nature of "Newtonian philosophy" in the eighteenth century, see Cohen, *Franklin and Newton*, pp. 180-1.
[51] *Philosophiae Newtonianae institutiones,* "Ad lectorem."
[52] "Histoire," p. XXXVI, n. T.
[53] *Introductio,* p. 13.
[54] For studies of "hypotheses" in Newton's own thought, see Cohen, *Franklin and Newton,* p. 129 ff. and Appendix I; Alexandre Koyré, *Newtonian Studies* (London: Chapman and Hall, 1965), Ch. II; and E. W. Strong, "Hypotheses non fingo," in *Men and Moments in the History of Science,* ed. Herbert M. Evans (Seattle: University of Washington Press, 1959), pp. 162-76.
[55] Cohen, *op. cit.,* p. 181.

taught his students about material particles of light and fire and attributed the effects of electricity to subtle but corporeal vapors.[56] Both Musschenbroek and 's Gravesande made much of the many "forces" that Newton had posited throughout nature and clearly affirmed with Newton the existence, though as yet undetectable, of absolute time, space and motion.[57]

Indeed, the significance of the frequent repudiation of hypotheses often appears to have resided less in its scientific meaningfulness than in its prominence in the polemic with the hard-line mechanists, above all, the Cartesians. In response to protests against the incomprehensibility of Newtonian "attractions" in nature, Musschenbroek retorted that at least they did not derive from mere hypotheses, "such as that playful subtle matter of Descartes, found nowhere but in the mind of a dreamer...."[58] Allamand likewise, in orienting his remarks on hypotheses towards "Descartes and his disciples," also reflected the importance of the debate with the Cartesians that remained always in the background.

Nonetheless, though partisan strife surely affected the way in which the Newtonians saw and represented themselves, Newton's *"hypotheses non fingo"* had a far deeper significance in their natural philosophy than its polemical convenience. The corpuscular imagery so critical to the Cartesian "hypotheses" played now, in truth, but a peripheral role in the physics of Musschenbroek and 's Gravesande. No longer did their physics consist of a progressive elaboration of that imagery, as for instance in the case of de Raey, nor was that imagery the basis, as with Senguerdius, of the critical rationale behind the design and interpretation of the bulk of their experiments. Despite the continuing strength of the seventeenth-century mechanistic tradition, the natural philosophy of Musschenbroek and 's Gravesande ultimately rested on concepts the mechanical inexplicability of which was recognized and accepted. Consequently, the rejection of hypotheses embodied a conviction that, for the

---

[56] *Introductio*, pp. 13, 254-5 and *passim*; *Elementa*, pp. 172, 309, 325-6 and 359. In advancing the materiality of fire, Musschenbroek was following the lead of Boerhaave. For the importance of the speculation on fire by Boerhaave, Musschenbroek and 's Gravesande in the development of the eighteenth-century doctrine of imponderable fluids, at least in Britain, see Schofield, *Mechanism and Materialism*, Ch. VII.

[57] 's Gravesande, *Philosophiae Newtonianae institutiones*, pp. 14, 80 and 294. Musschenbroek, *Elementa*, pp. 52-5. On forces in the works of 's Gravesande and Musschenbroek, see below, p. 124 ff.

[58] "Oratio de methodo instituendi experimenta physica," in Musschenbroek's *Tentamina experimentorum naturalium captorum in Academia del Cimento* (Lugduni Batavorum: Apud Joan. et Herm. Verbeek, 1731), p. XXXIV.

sake of science itself, the inability of human reason to achieve a complete understanding of the workings of nature had to be acknowledged. It was a call to abandon the high hopes of Cartesian rationalism, and, as such, it was a significant hallmark of the new science as it emerged from the hands of Newton.

Even with the assistance of the senses, the necessity of which both Musschenbroek and 's Gravesande reaffirmed,[59] reason could not penetrate, they maintained, to a knowledge of the inner foundation of physical being, to a comprehension of substance. In the preface to his *Physices elementa mathematica*, 's Gravesande again echoed Newton in asserting that substance remained completely hidden and that matter might have properties of which as yet they had no idea.[60] Musschenbroek likewise warned his students at the beginning of his course that, since we learn through the senses only what pertains to the surface of bodies, many things continued unknown to us, including the inner content of material being: "What truly is that which remains confined within the surface of bodies? Is it not that which specifically (*proprie*) forms the body and is its substance? But of this, we are ignorant." [61] And of things unknown to us, admonished 's Gravesande, we are to deny or affirm nothing.[62]

The most notorious instance of the Newtonian betrayal of the quest for complete comprehension in the philosophy of nature was universal gravitation, the essential innovation on which Newton had rested his mathematical analysis of the solar system. Together with the laws of motion, wrote 's Gravesande, universal gravitation laid bare "the whole artifice by which the prodigious machine, the system of the planets, is ruled." [63] The planets moved now through an infinite space which, if not completely empty, contained only a celestial matter that did not affect the motion of the planets in any perceptible way.[64] Though moving in a virtual vacuum, the planets were held in their elliptical orbits about the sun by forces that bore the sun and planets towards each other, the center of the solar system now being the common center of gravity of all the member bodies taken together. The gravitational attraction that so governed the solar system was identical with the gravity that affected terrestrial bodies as well, and, hence, though 's Gravesande acknowledged

---

[59] 's Gravesande, *Physices elementa mathematica*, Vol. I, "Praefatio." Musschenbroek, *Introductio*, pp. 4 and 23.
[60] *Loc. cit.*
[61] *Elementa*, p. 11.
[62] *Physices elementa mathematica*, Vol. I, "Praefatio."
[63] *Philosophiae Newtonianae institutiones*, p. 333.
[64] *Ibid.*, pp. 6, 285 and 345. For 's Gravesande, the vacuum did indeed exist; *Ibid.*, pp. 4 ff. and 342 ff.

that the bodies of the universe beyond the solar system were too distant for their motions to be observed,[65] the motions of at least the immediate heavens shared with terrestrial phenomena a common derivation – and, consequently, a common inexplicability.

Both Musschenbroek and 's Gravesande stressed that gravitation could not be understood in terms of any comprehensible mechanics. If there was a matter which, by driving against bodies, produced the effect of gravity, asserted 's Gravesande, it had to be fluid, and a fluid so subtle that it freely penetrated the pores of all bodies, since one body enclosed within another body still remained heavy, and so rare that it did not perceptibly impede the motion of bodies, since the motion of a pendulum continued at great length within a vacuum.[66] He challenged any mathematician to demonstrate how the phenomena of gravity could be brought about by so rare and subtle a fluid, and how such a fluid could account for the fact that the effect of gravity was exactly proportional to the quantity of matter in a body rather than to surface area, against which alone a fluid could press. "We are not denying," he concluded, "that gravitation derives from impact. But we do contend that gravitation obviously *does not follow from any impact according to the laws we know,* and we admit its cause to be completely concealed from us." [67] Dropping even 's Gravesande's reservation that there might perhaps be impact according to laws that were as yet unknown, Musschenbroek not only rejected a mechanistic explanation for gravity but stressed as well the futility of considering any other kind of causation:

Or does gravity depend, rather, upon a certain universal spirit which penetrates not only pores but corporeal substance itself? Whether such a spirit is to be found in the universe we do not know, nor do we understand how non-corporeal spirit works within body. Is there not then some active internal principle of gravity by the force of which terrestrial bodies are bound to the center of the earth? ... Of this, however, we can form no clear idea in our minds, since we are not permitted to peer into the internal substance of bodies. Nor do we understand how or why heavy bodies at a distance from each other mutually affect each other with nothing in between. Is it then an essential attribute of bodies? This I do not affirm, since I am ignorant of their essence....[68]

Gravitation was an "attraction," and 's Gravesande emphasized, as had Newton, that this designated only an effect, two bodies gravitating or tending towards one another, a phenomenon, not a cause.[69] The micro-

---

[65] *Ibid.,* p. 286.
[66] *Physices elementa mathematica,* Vol. II, pp. 152-3.
[67] *Ibid.*
[68] *Elementa,* pp. 98-9; see also pp. 42-5, 82-3 and 99-102.
[69] *Philosophiae Newtonianae institutiones,* p. 11.

scopic as well as the macrocosmic Newtonian world was structured in fact by a host of such "attractions" (and repulsions) of which gravitation was only the most eminent. 's Gravesande dwelt at length on refraction as a consequence of light's attraction to body, and he depicted fire as well, about which he otherwise explicitly declined to feign hypotheses and left quite ill-defined, as also attracted to bodies and inducing mutual forces of repulsion among their particles.[70] The elasticity of the air he attributed to a similar repulsion among its particles, while hardness, softness, fluidity and elasticity in general were functions of corpuscular "cohesion," the result, again, of an "attractive force" the cause of which was unknown.[71] A wide spectrum of other phenomena, from capillarity to chemical effects and magnetism, were cited by Musschenbroek and 's Gravesande as further examples of the existence of attractions and repulsions at work in nature,[72] the causes of which were all as unintelligible as the cause of gravitation.

This incomprehensibility penetrated still more deeply into the physics of Musschenbroek and 's Gravesande, for it extended to the causality of motion in general, the importance of which they both had reaffirmed. "All things that take place in natural objects," asserted 's Gravesande, "pertain to motion," and Musschenbroek declared motion to be the principal object of physics.[73] 's Gravesande defined it as a "passage (*translatio*) from place to place, or a continual change of place," while Musschenbroek rendered it "the passage of a body from one part of space to another." [74] The motion they conceived was now thoroughly inertial, but, as Musschenbroek's definition suggests, they followed Newton in reasserting the reality of absolute motion, even though it was undetectable. Space – "since it is of infinite extent," said Musschenbroek – was immobile, and hence the existence of absolute as well as relative "place" and absolute as well as relative motion.[75] But only relative motion could be recognized by the senses, acknowledged 's Gravesande, for "who could reasonably either affirm or deny that all the bodies known to us are not being borne through the immensity of space with a common motion?" [76]

---

[70] *Ibid.,* pp. 199 ff., 185-6 and 195. 's Gravesande conceived of fire as very closely related to light, for it was fire entering the eye in straight lines that produced the impression of light; *Ibid.,* p. 191.

[71] *Ibid.,* pp. 161 and 9-11.

[72] 's Gravesande, *Physices elementa mathematica,* Vol. I, pp. 9-13. Musschenbroek, *Elementa,* pp. 185 ff. and 196 ff.

[73] *Physices elementa mathematica,* Vol. I, p. 14. *Elementa,* pp. 3-4.

[74] *Ibid.*

[75] Musschenbroek, *Elementa,* pp. 40 and 50. 's Gravesande, *Philosophiae Newtonianae institutiones,* pp. 6 and 14.

[76] *Ibid.,* p. 294.

Perhaps misled by the difficulties and ambiguities confronted in Newton's thought, his disciples at Leiden also clung to older sensibilities in continuing to assume that all motion, including inertial motion, required a cause.[77] "Because a body is moved," taught Musschenbroek, "it is conveyed from one part of space to another; this transfer is a real effect, which needs a real cause in the body. This cause is a force (*vis*) driving the body." [78] "Force," indeed, in the Newtonian physics taught by 's Gravesande and Musschenbroek had become a generalized cause behind all motion. "I call *force*," wrote 's Gravesande for a more professional audience than his classroom, "that which, in a moving body, transports it from one place to another." [79] As the causality behind all motion, force had become, in effect, a philosophical concept prior even to the concept of motion itself.

Echoed in the statements of Musschenbroek and 's Gravesande was Newton's identification of force in one capacity, as *vis insita,* with inertia and its unchanging perseverance. Ultimately, however, the concept of force was to be limited more strictly to a contrary effect for which it was also responsible in Newtonian physics, the altering of motion, or change in its direction or velocity. Of force in this second aspect, gravity and other attractions and repulsions were, in effect, pure embodiments, pure forces working independently of any detectable mechanism. Gravity, taught Musschenbroek, was "a force by which terrestrial bodies in the open air or freed within a vacuum (*vacul*) are borne downwards from rest in a line perpendicular to the horizon of the earth." [80] The "cause" of attraction in general, which he otherwise declared to be unknown, was also to be called simply an "attracting force." [81] 's Gravesande expressed himself likewise: "The force by which bodies are driven towards the earth is called *gravity*"; "We understand the word *attraction* to mean *any force by which two bodies tend towards one another.*" [82] On the one hand the generalized cause of motion, force was also manifested in a variety of pure embodiments in nature that, together with the force that caused inertia, organized and activated the universe.[83]

[77] For the meaning and development of Newton's own conception of force, see Richard S. Westfall, *Force in Newton's Physics: The Science of Dynamics in the Seventeenth Century* (London: Macdonald; New York: American Elsevier; 1971), Ch. VII and VIII.
[78] *Elementa,* p. 56.
[79] "Essai d'une nouvelle theorie du choc des corps, fondée sur l'experience," in *Oeuvres philosophiques et mathématiques,* Vol. I, p. 219.
[80] *Elementa,* p. 82.
[81] *Ibid.,* pp. 182-5.
[82] *Philosophiae Newtonianae institutiones,* pp. 18 and 11.
[83] See Musschenbroek, *Introductio,* pp. 50-1.

Despite its pervasive and fundamental role in their natural philosophy, however, both 's Gravesande and Musschenbroek neglected to provide a definition of force as such, *vis,* within their texts. Indeed, though now basic to continental as well as British schools of thought, the concept of force remained a profound and troubling source of confusion for the European scientific community at large. It could not be denied, 's Gravesande had written to his scientific peers, that force – once again primarily in its inertial capacity, however – was something real (*positif*), since it was the cause of a sensible effect.[84] But what kind of real thing it was he did not attempt to say, and he acknowledged to his students that the cause of the continuation of inertial motion – which would be force in that same inertial capacity – appeared completely unknown to him.[85]

Musschenbroek, characteristically, was less laconic about the unknowns and the paradoxes in his natural philosophy. He had also referred to force as a "real cause" that passed from body to body, penetrating to their innermost parts, and "not through pores but through solid substance itself...."[86] He had continued, however:

Is force then physical being? Or a unique substance? Or is it an idea first produced in the perceiving mind and thereafter imparted to bodies, among which it passes from one into another? None of these can be demonstrated, and it is better to admit our ignorance, or acknowledge that our minds are not fit to form a clear idea of this thing.[87]

The obscurity of force now undermined even the seeming comprehensibility of interaction through contact, as Musschenbroek did not fail to point out to those still complaining of the incomprehensibility of attraction. We do not know what force is, he declared, nor how it inheres in bodies, moves them, or passes from one into the other.[88] We see and understand nothing beyond the commonplace effects, and "hence men are equally as blind with respect to external principles of operation as internal." As an unintelligible but nonetheless "real cause" of all motions and actions in nature, whether effected through contact or at a distance, the concept of force as taught in the Newtonian physics of Musschenbroek established an ultimate incomprehensibility behind all physical causality, the understanding of which, historically, had been the goal of natural philosophy.

Musschenbroek and 's Gravesande continued to speak of physics,

---

[84] "Essai d'une nouvelle theorie du choc," p. 219.
[85] *Physices elementa mathematica,* Vol. I, p. 36.
[86] *Elementa,* p. 56.
[87] *Ibid.,* p. 57.
[88] *Ibid.,* p. 185.

nonetheless, as an inquiry into the causes in nature. The perfection of physics, wrote Musschenbroek, required more than a concern with the mathematical relations and consequences of phenomena, it demanded a knowledge of causes as well.[89] "Physics," 's Gravesande had declared in his inaugural address at Leiden, "teaches us what the causes of natural phenomena are...." [90]

With so many and such fundamental "causes" acknowledged as unknown or incomprehensible, to continue to charge physics with the traditional task of making known the causes in nature might seem somewhat perverse. But 's Gravesande had been more explicit about his meaning: "Physics teaches us what the causes of natural phenomena are; that is to say, it seeks out what the laws are to which the Creator found it fitting to submit the universe, and to which He desired motions that bring about all changes there to conform." [91] The aim of physics, he wrote elsewhere, was to find the laws of nature from which the phenomena were to be derived, and the discovery of these laws was perhaps as far as one could go in the search for causes.[92] As taught by 's Gravesande and Musschenbroek at eighteenth-century Leiden, the laws of nature were advanced as causal explanations in a physics in which the meaningfulness of natural causality was otherwise rapidly fading. But the causality of these laws was itself problematic.

's Gravesande defined a natural law at the beginning of his course as a "rule and pattern according to which God desired certain motion always, on all occasions that is, to proceed." [93] Also affirming their origin in the will of God, Musschenbroek went on to stress their apparent arbitrariness as well.[94] God could have established the laws of nature quite otherwise, and why he wanted them as they were was beyond comprehension. These laws followed, consequently, from no reason or rationale the mind of man could discern: "Such laws are learned only from the observation of the senses, and the wisest of mortals has not been able to discover any one of them through contemplation alone...." [95]

Nor did the laws of nature follow from any physical necessity that was as yet understood. Whether they derived from the essence of matter, or from some nonessential properties that God had nonetheless imparted to bodies, or from causes quite alien to our thinking was uncertain,

[89] *Essai de physique,* Vol. I, "Preface."
[90] "Discours sur l'utilité des mathématiques," p. 319.
[91] *Ibid.*
[92] *Physices elementa mathematica,* Vol. I, "Praefatio."
[93] *Philosophiae Newtonianae institutiones,* p. 2.
[94] *Elementa,* p. 5.
[95] *Ibid.,* pp. 4-5.

's Gravesande confessed, and he included in his broader definition of the laws of nature that their own causes were unknown.[96] "It is enough for us to detect the manner in which they have been established," added Musschenbroek, 'and to admire the supreme wisdom of the Creator in the elegant order of the universe. The cause and reason of these laws, accordingly, is unknown to us." [97]

Musschenbroek and 's Gravesande were again speaking of "causes" whose own causes, short of God, were incomprehensible. The laws of nature, however, differed significantly from such causes as gravity, attraction, repulsion or force in general. In the physics of Musschenbroek and 's Gravesande, the latter causes still suggested real, even if unintelligible, physical agencies, literally compelling each individual phenomenon to occur as it did. The laws of nature, on the other hand, were in reality little more than recognized consistencies in phenomena. "We call *laws*," wrote Musschenbroek, "constant appearances (*apparitiones*), which, however often bodies are placed in similar situations (*in similibus occasionibus*), always occur in the same way." [98] 's Gravesande likewise further explained the laws of nature "with respect to us" – a phrase that seemed to have meant a limitation to that purely secular knowledge founded only on man's experience of the physical world – as "every effect which is the same on all occasions...." [99]

Near the end of his life, Musschenbroek summarized the idea of the laws of nature as it had been taught at Leiden for almost half a century:

And thus, with respect to us, laws are simple effects which are the same on all similar occasions and for which, though they may perhaps follow from other simpler or more general laws, no other law from which they follow as from a cause is perceived. The laws of which we speak, to be sure, do not reveal whether a thing derives from the will of God directly, or is brought about by an immediately preceding but unknown intermediate cause, or by a long series of other causes. It can hardly be doubted, indeed, that some laws are primary laws, proceeding immediately from the will of God, while others are secondary, deriving, in turn, from the primary laws.

But all laws are unfailingly constant, for God is always the same being, absolutely perfect, wise and immutable, and the divine will and providence, therefore, is unfailingly constant and flawless.

These laws can be invoked as often as similar phenomena occur, and in this

[96] *Physices elementa mathematica,* Vol. I, "Praefatio." *Philosophiae Newtonianae institutiones,* p. 2.
[97] *Elementa,* p. 5.
[98] *Ibid.,* p. 4. Musschenbroek also defined what he meant by "appearances," *i.e.,* phenomena: "All positions, motions, changes, and actions of bodies which are observed by the senses, either by one of these or by many, are called *phaenomena* or *apparitiones*"; *ibid.,* p. 3.
[99] *Philosophiae Newtonianae institutiones,* p. 2.

way the laws are illustrated and confirmed by examples. Since their causes are hidden, the philosopher can scarcely advance any further with certainty.[100]

The significance of these laws was but a reliable consistency in nature the cause of which remained, philosophically at least, obscure. The discovery of prior or more general laws from which those already known might be deduced only removed the ignorance of causality to but a broader level. As natural philosophy extended its knowledge of the laws of nature – and that many such laws were yet to be discovered, asserted 's Gravesande, anyone who had studied the phenomena of nature would be convinced [101] – it progressed through a succession of laws the most general of which always remained rooted, as it were, on the other side of the frontier between the known and the unknown. Musschenbroek and 's Gravesande were certain only of the ultimate, divine source of these laws, but even then, the why and the wherefore of the divine ordinances lay beyond mortal understanding, and the inability to conceive how spiritual substance could move or affect corporeal substance remained, as Musschenbroek had testified,[102] one of the enduring legacies of the mechanical philosophy. As "causes," therefore, the laws of nature were merely observed consistency cited as an explanation of the individual instance, but for more than this the natural philosopher was now told he could scarcely hope.

These laws had emerged as the successors to the "first principles" of the preceding systems of natural philosophy. In truth, neither Musschenbroek nor 's Gravesande were now much concerned with the problem of philosophical first principles, nor did they give as much conscious attention as had their seventeenth-century predecessors to the organization or system that obtained among the concepts they used. Throughout the greater part of their physics, however, it was on the laws of nature that they rested their reasoning. It was through the three Newtonian laws of motion that motion and force acquired an applicable meaning, and it was as a law that gravitation, together with the laws of motion, tied together the Newtonian solar system that 's Gravesande, if not Musschenbroek, explained at considerable length.

In cultivating a physics founded on laws the causes of which were accepted as unknown, Musschenbroek and 's Gravesande abandoned to that extent the traditional purpose of physics as a science, to understand nature through an understanding of her causes. Burgersdijck had affirmed that "science" in the strictest sense was "knowledge of necessary

[100] *Introductio*, pp. 7-8.
[101] *Physices elementa mathematica*, Vol. I, "Praefatio."
[102] See above, p. 124.

things through their causes," [103] and the Cartesians had embraced this Aristotelian conception of science with even greater ardor than their scholastic predecessors. De Volder held physics to be a search for the causes of phenomena and identified that search with the question of first principles that had so preoccupied him: "What is more evident," he asked, "than the general truth that the cause of no phenomenon can be rightly explained without the knowledge of principles or first causes?" [104] The late scholastics, including Burgersdijck, generally acknowledged that their knowledge of causes in physics often lacked the clarity and certainty that their ideal of science would demand,[105] but the Cartesians no longer tolerated such imperfection. Convinced that their principles were absolutely certain and absolutely comprehensible, they promised what was in effect the final fulfillment in physics of the traditional ideal of science, and it was a promise that had inspired professors and students at Leiden for well over half a century. But 's Gravesande and especially Musschenbroek, despite their frequent but loose use of the word "cause," had now acquiesced in the ultimate incomprehensibility of physical causality.

Musschenbroek and 's Gravesande had not depicted their physics, however, as being distinguished only by the rejection of hypotheses and the admission of ignorance. They characterized it as well as a science founded on mathematics and experimentation. 's Gravesande's major work, as we have seen, was entitled "The Mathematical Elements of Physics Confirmed by Experiments," and critics in France and Germany even protested that he seemed to be implying that the recourse to geometry and experiments was distinctively Newtonian alone.[106] Musschenbroek declared in the preface to the *Elementa physicae* that the passion for inventing hypotheses had at last been bridled, "and in their place have come proper geometric demonstrations, careful observations, and purposeful experiments...." [107] The consequence was a "true and stable method of philosophizing" that could, after all, establish "truth and certainty" in physics.

In phrases that must by now have seemed only too familiar, 's Gravesande had declared in his inaugural address that mathematics offered the only foundation for a true physics.[108] The mathematical foundation he

---

[103] *Collegium physicum*, p. 2.
[104] *Disputationes philosophicae de rerum naturalium principiis*, pp. 4 and 12.
[105] Reif, "Natural Philosophy," pp. 36, 42, 264 and 290 ff. Burgersdijck, *Collegium physicum*, p. 3. See also Kyper, *Institutiones physicae*, p. 5.
[106] See Allamand, "Histoire," pp. XXXVI, n. T., and XXXVII-XXXVIII, n. U.
[107] *Elementa*, "Praefatio."
[108] "Discours sur l'utilité des mathématiques," p. 320.

envisaged, however, was clearly no longer the unsullied rationalism sought by his Cartesian predecessors. He depicted all mathematical reasoning as pertaining to the comparison of quantities and, in natural philosophy, resting on the laws of motion – indeed, it was precisely because motion was a quantity, he argued, that the whole of physics was to be treated mathematically.[109] Accordingly, the texts of both Musschenbroek and 's Gravesande display a succession of geometrical representations of flying projectiles, spouting fluids, orbiting planets, and bending light. Burgersdijck would doubtless have maintained a century earlier that the subject they were teaching was not physics but mathematics, and 's Gravesande himself acknowledged that physics might now reasonably be considered a branch of mathematics.[110]

Implicit in 's Gravesande's remarks was the fact that the laws of motion were themselves mathematical. Moreover, he had explicitly confirmed that these laws were to be learned only through the examination of phenomena, an undertaking that, since nature was so secretive, often necessitated the employment of "art."[111] Musschenbroek similarly observed that the method for acquiring knowledge of forces and their quantities was through "evoking and observing phenomena, which are provided by bodies either spontaneously or through planned experiments."[112] The mathematics of measurement and comparison now joined the science of the laws of nature to experimentation, which Musschenbroek in his turn maintained was the only basis of reasoning in physics.[113]

Though acknowledging that there were those who considered experiments to be but childish games, Musschenbroek exalted experimental physics as "the furthest limit, the highest peak, of science," the attainment of which could lead to immortal glory,[114] and both he and 's Gravesande succeeded in winning for themselves a distinguished place, if not quite immortal glory, in the history of eighteenth-century experimentation. Musschenbroek's most celebrated experimental endeavor was that involving the discovery of the Leyden jar, which rapidly became an essential instrument in the eighteenth century's preoccupation with the phenomena of electricity.[115] In late 1745 and early 1746, Musschenbroek,

[109] *Ibid. Physices elementa mathematica,* Vol. I, "Praefatio."
[110] *Physices elementa mathematica, loc. cit.*
[111] *Ibid.*
[112] "Oratio de methodo instituendi experimenta physica," p. VIII.
[113] *Ibid.,* p. X.
[114] *Ibid.*
[115] See Park Benjamin, *The Intellectual Rise in Electricity* (New York: D. Appleton and Company, 1895), p. 511 ff. Herbert Dingle, "Physics in the Eighteenth Century," in *Natural Philosophy Through the 18th Century and Allied*

Allamand, and a wealthy citizen of Leiden named Cunaeus were experimenting together with one of the static electricity machines which by then had become popular throughout Europe. From a rapidly revolving glass globe that, when brushing against an experimenter's hands, generated electricity, a contact was connected to a wire the far end of which was suspended in a jar partly filled with water. At the end of an experiment in which Musschenbroek was holding the jar, he touched the wire to remove it from the water and received what he described as "a shock of such violence that my whole body was shaken as by a lightning bolt." [116] Musschenbroek's stirring account of the event was read before the *Académie des sciences* in Paris and soon published, enhancing Musschenbroek's reputation and bestowing upon the new discovery – though in fact it had already been stumbled upon earlier in 1745 in Pomerania – the name it still bears.

More exemplary, even if less fruitful, of the experimental science Musschenbroek and 's Gravesande were urging upon their students was a series of experiments on magnetism that Musschenbroek had undertaken while still at Utrecht. His primary purpose had been to discover some proportion between the effect two magnets had upon each other and the distance that separated them.[117] To accomplish this, he had suspended one magnet above another by a long thread connected to a distant balance, whereon he could weigh the "quantity of forces" exerted between the two magnets. During the course of these same experiments, he also sought evidence for what he admitted to have long been his belief, that the effect of magnetism was the consequence of some corporeal fluid or effluvium. Believing "nothing to be more injurious to the advance of science than swallowing conjectures in the place of demonstrations," he endeavored to confirm his belief by interposing large masses of lead, tin, silver, copper and mercury between the two magnets in anticipation of thereby obstructing whatever was responsible for their attraction. Not even a diminution of the forces between the two magnets was detected, however, which he found to be quite astonishing and "beyond the understanding of any mortal." He concluded that the cause of magnetism was not only perhaps noncorporeal but completely unknown. His efforts to find a proportion between distance and the force of magnetism was unsuccessful as well, for he was compelled to conclude that no such proportion appeared in his findings. Nonetheless, it was a

---

Topics, ed. Allan Ferguson (a commemorative number of *The Philosophical Magazine*; London: Taylor and Francis, Ltd., 1948), pp. 39-41.

[116] Benjamin, *op. cit.*, p. 519.

[117] "De viribus magneticis," in the *Philosophical Transactions*, Vol. XXXIII (for the years 1724-5; published 1726), pp. 370-7.

problem that would prove of great importance in the subsequent progress of modern physics, and Musschenbroek's experiments underscored the fact that it was a problem of measuring and finding mathematical proportions in that which remained incomprehensible.

A still more sophisticated example of such experimentation was offered in the innumerable experiments performed by 's Gravesande in his continuing attempt to resolve the eighteenth-century controversy as to whether the force to be attributed to a moving body varied according to its velocity or the square of its velocity.[118] Convinced by experiments he performed during his early years at Leiden, 's Gravesande published in the *Journal littéraire* in 1722 an "Essai d'une nouvelle théorie sur le choc des corps" [119] in which, though acknowledging the ambiguity in the general usage of the word "force," he emerged as a spokesman for the square of the velocity. In so doing, he abandoned the British Newtonians on this issue for the position championed chiefly by the followers of Leibniz, and he aroused no little ill will, thereby, across the Channel.[120] He remained, however, one of the more moderate as well as one of the more perspicacious participants in the controversy, and continued to rest his arguments entirely on careful and repeated experiments. Most decisive in shaping his own views had been a series of experiments in which bodies of the same shape but of weights differing according to a simple proportion were dropped from heights which also differed in a simple proportion.[121] The bodies fell onto a plate of clay, and the impressions they left were measured and correlated with the weights of the bodies and the heights from which they had fallen. Designing his own elaborate instruments, 's Gravesande also carried out numerous experiments to determine the effects of impact between colliding bodies, both elastic and inelastic. Many of his experiments were not of his own original devising, but he performed them with a precision and skill that won wide recognition.

It was through their classroom teaching, however, and what were in effect introductory textbooks that 's Gravesande and Musschenbroek

---

[118] On this dispute and 's Gravesande's part in it, see Thomas L. Hankins, "Eighteenth-Century Attempts to Resolve the *Vis viva* Controversy," *Isis*, Vol. LVI (1965), pp. 281-97; Pierre Costabel, "'s Gravesande et les forces vives ou des vicissitudes d'une expérience soi-disant cruciale," in *Mélanges Alexandre Koyré*, Vol. I: *L'Aventure de la science* (Paris: Hermann, c. 1964), pp. 117-34; and L. L. Laudan, "The *Vis viva* Controversy, a Post-Mortem," *Isis*, Vol. LIX (1968), pp. 131-43.

[119] See his *Oeuvres philosophiques et mathématiques*, Vol. I.

[120] Allamand, "Histoire," p. XXXIV, n. T. Hankins "Eighteenth-Century Attempts," p. 289.

[121] Hankins, *op. cit.*, p. 287.

exercised their greatest influence in the eighteenth century, and, as de Volder had observed in passing, pedagogy and active research had their different demands. In a pedagogical context still oriented towards young "beginners," [122] experimentation was no longer directed towards investigation and discovery but, like earlier scholastic "methods," towards the clear and easy presentation of what was already known. The importance of experimentation as a teaching devise was further emphasized by the recognition of how little in the way of mathematics the students would learn in the traditional university curriculum. Even 's Gravesande only aspired as professor of mathematics to induce the students to give "a little portion" of the time they passed at the university to the study of geometry and algebra, and the first edition of the *Physices elementa mathematica* had those who had acquired only the rudiments of mathematics very much in mind.[123] Characterized by Boerhaave's renowned pupil Albrecht von Haller as a skillful but somewhat ineloquent mathematician, 's Gravesande himself expressed his intention in the *Physices* to render the study of physics easier and more pleasurable by illustrating mathematical conclusions with experimental demonstrations, a "method of teaching natural philosophy" which he acknowledged as having been learned from the English.[124]

Haller also found it worth remarking that 's Gravesande had succeeded in constructing from Newton's thought a complete framework (*Gestelle*), presumably meaning the framework of the physics 's Gravesande taught, the most outstanding aspect of which Haller also found to be the experimental demonstrations.[125] Despite the fact that 's Gravesande showed less interest in clarifying the systematic coherence that tied together his philosophic concepts, he shared with his seventeenth-century predecessors their attentiveness to systematic coherence in pedagogy, and the very procedure of experimental – and mathematical – demonstration became for him a fundamental source of that coherence in the discipline of physics. Even in his third and last edition of the *Physices*, the edition most oriented towards his scientific peers, he still spoke of his decisive concern for shaping mathematical and experimental demonstrations into a

---

[122] Musschenbroek intended the some 243 experiments in the *Physicae experimentales et geometricae dissertationes*, most of which had been demonstrated in the university auditorium at Utrecht, for *usus tironum*; "Praefatio."

[123] "Discours sur l'utilité des mathématiques," p. 315. See the *Physices elementa mathematica*, Vol. I, "Praefatio," and Vol. II, "Lectori S."

[124] *Haller in Holland: Het Dagboek van Albrecht von Haller van zijn Verblijf in Holland* (1725-1727), ed. G. A. Lindeboom (Delft: Koninklijke Nederlandsche Gist- en Spiritusfabriek N.V., 1958), p. 37. 's Gravesande, *Physices elementa mathematica*, Vol. I, "Praefatio."

[125] *Haller in Holland*, p. 94.

"system," [126] and the *Physices* itself was in fact the first general text to attempt to present Newtonian physics primarily through an elaborate series of experiments.[127] Experimentation had first been introduced at Leiden by de Volder as a pedagogical novelty of uncertain philosophic merit; now dominating the pedagogy of the discipline, experimental activity appeared all the more vividly to the young students at Leiden as an essential feature of natural philosophy.

In admiring 's Gravesande's many experimental demonstrations, Haller had been particularly struck by the Dutch professor's impressive instruments.[128] In collaboration with Musschenbroek's brother, Jan, 's Gravesande was in fact creating one of the century's finest collections of experimental equipment.[129] Numbering well over two hundred items at 's Gravesande's death, his personal collection was purchased for the university by the curators, who then had to enlarge the auditorium for experimental physics in order to house the new acquisitions.[130] Musschenbroek also acquired a large personal collection for both research and teaching, but that collection, unfortunately, was dispersed at his death by auction.[131]

The new machines designed and collected at Leiden amply illustrated the increasing ingenuity with which nature was being vexed and interrogated. The air pump still remained a popular instrument of inquisition, and 's Gravesande, like Senguerdius, designed his own and produced a glass model to reveal to his students its inner operation.[132] (The air pump was an instrument for distressing nature in more than a figurative sense, for the experiments of both Senguerdius and 's Gravesande continued to entail the suffering and death of small animals.)[133] But 's Gravesande now marshaled a host of other new and elaborate machines as well. Centrifugal forces were measured and compared for the students by means of a large table with three interconnected revolving discs, and the

---

[126] Costabel, "'s Gravesande et les forces vives," pp. 124-5.
[127] Schofield, *Mechanism and Materialism*, p. 141.
[128] *Haller in Holland*, p. 94.
[129] Crommelin, "Physics and the Art of Instrument Making," p. 35.
[130] *Bronnen*, Vol. V, pp. 137*-42*, 246 and 268. 's Gravesande had also added to the university's own collection of scientific instruments as well, including its collection for astronomy; see *Bronnen*, Vol. V, pp. 136*-7* and 144*-6*, and also de Sitter, *Short History of the Observatory*, p. 16. Many of the instruments from 's Gravesande's personal collection today form the nucleus of the collection of the Museum of the History of the Natural Sciences in Leiden.
[131] Crommelin, "Physics and the Art of Instrument Making," p. 37.
[132] *Physices elementa mathematica*, Vol. I, pp. 161 and 172, plates XXVII and XXXII. Crommelin, *Instrumentmakerskunst*, pp. 14-5.
[133] *Physices elementa mathematica*, Vol. I, pp. 169-70, plate XXXI. Senguerdius, *Inquisitiones experimentales*, p. 13 ff.

laws of impact were similarly demonstrated by a machine designed by 's Gravesande and subsequently improved by Musschenbroek.[134] Fluids were forced upwards by varying weights on a compressible drum or were spouted in arched jets down long troughs, into which, at other times, would be suspended a long pendulum, its bob immersed in water in order to investigate fluid resistance.[135] Other devices included large waterfilled prisms, a static electricity generator, a complex machine for striking sparks in a vacuum, and a small toy-like steam vehicle.[136]

The science of nature as portrayed in the discipline of physics at Leiden was now deeply colored by an increased interest in machines themselves. A discussion of the simple machines, including the balance, the lever, the pulley and tackle, the wheel and axle, the inclined plane, the wedge, the cogwheel and the threaded screw, now appeared among the topics with which Musschenbroek and 's Gravesande began their courses. 's Gravesande devoted special chapters in his textbooks to the study of the pendulum, the air pump, the microscope and telescope, mirrors, and the magic lantern, and Musschenbroek introduced a chapter on friction in machines. By contrast, the realm of living things, so important in the physics of Burgersdijck, seldom appeared in the teachings of his eighteenth-century successors. Physics at Leiden was now a discipline that revolved primarily aroung mechanics, fluid mechanics, optics and, in 's Gravesande's classroom, astronomy.[137]

The increased importance of machines in the discipline of physics at Leiden, however, did not reflect a continuing predisposition to perceive nature herself as but a machine differing from man-made contrivances only in size and complexity. The operation of machines was now "natural," as it had not been for Burgersdijck, but the workings of nature herself far transcended such operation. Musschenbroek and 's Gravesande did not suggest, as had Senguerdius, that their instruments duplicated nature's methods, and they vigorously denied that the mechanistic imagery of the Cartesians was capable of representing nature's most fundamental processes. 's Gravesande spoke of the solar system, indeed,

---

[134] *Physices elementa mathematica*, Vol. I, pp. 77-9, plate XVI. Musschenbroek, *Introductio*, p. 251.

[135] *Physices elementa mathematica*, Vol. I, pp. 105, 133 and 124-5, plates XX, XXIII, XXIV and XXII.

[136] *Ibid.*, Vol. II, pp. 90, 3, 13 and 16, plates XVII, I, II and III.

[137] Musschenbroek gave far less attention to astronomy in his texts than did 's Gravesande, but, on the other hand, he included a consideration of meteorology. Like Senguerdius before him, Musschenbroek carried on his own systematic studies of meteorology; see the concluding passages and the graph in the *Physicae experimentales et geometricae dissertationes*, and Kuenen, *Het Aandeel van Nederland*, pp. 113-4.

as a "prodigious machine," but his recognition of the failings of the mechanical philosophy implicitly entailed a recognition of the limited applicability of the analogy of the machine in natural philosophy. What did remain of the mechanical image of nature was the conviction that nature's workings, like those of a machine, were ultimately governed by an inescapable and predictable invariability, an invariability that, through the agency of machines, could be worked upon and further exposed.

In the physics now taught at Leiden, the aggressively inquisitive practice and attitudes of experimental science were confirmed as not only proper but central to the practice of natural philosophy. The continual example of experimental procedures in application, even if only for pedagogical demonstration, conveyed what could not be conveyed by mere admonishments, such as Heereboord's, that physics was the study of nature rather than of the books and opinions of men. There is no evidence that the students themselves performed experiments,[138] but 's Gravesande and Musschenbroek, like Senguerdius before them, were demonstrating not only the results but the "method of making experiments" as well.[139] It was a lesson that could be effectively taught by example alone, as was also the case with the application of mathematics. There could be, as we have seen, more than one conception of a mathematical physics, but Musschenbroek and 's Gravesande illustrated through their own procedures in the classroom how mathematics was being used by the builders of modern science.

While urging their students to curb the "excessive longing to understand," [140] Musschenbroek and 's Gravesande introduced them to the scientific practices whose success in leading to endless discovery and ever greater control over the phenomena of nature would increasingly obscure the philosophical ignorance to which the early Newtonians had had to reconcile themselves. The image of physics as a stable schematization of the world of nature was having to make way for a new image of an enterprise of ceaseless investigation, discovery, and change. Conse-

---

[138] J. A. Vollgraff, "Leidsche Hoogleeraren in de Natuurkunde in de 16de, 17de en 18de Eeuw," in *Jaarboekje voor Geschiedenis en Oudheidkunde van Leiden en Rijnland*, orgaan der Vereeniging "Oud-Leiden" (Leiden: A. W. Sijthoff, 1913), p. 190.

[139] On the title page to the *Rationis atque experientiae connubium*, Senguerdius explicitly stated that the work was intended to provide, among other things, a description of the method of making experiments, and Musschenbroek made that method the subject of his two major academic addresses, the "Oratio de methodo instituendi experimenta physica" and the "Oratio de certa methodo philosophiae experimentalis."

[140] 's Gravesande, *Physices elementa mathematica*, Vol. I, "Praefatio."

quently, the discipline of physics as now taught at Leiden served less as a preparation for the higher faculties in the schools than as an invitation to join in the restless scientific activity still carried on largely outside the universities. Whoever solves the riddles of electricity with which natural philosophy was now confronted, Musschenbroek tempted his students, would have his name "carved on monuments of eternal praise." [141]

---

[141] *Elementa*, p. 182.

CHAPTER VIII

CONCLUSION:

SCIENCE, PHILOSOPHY AND PEDAGOGY

In the early seventeenth century, the discipline of physics at Leiden, as elsewhere, had been nothing more than a part of an introductory scholastic preparation for more advanced study in medicine, theology or law. A century later, that discipline had become at Leiden an introduction into the concepts and procedures of modern science, a new and revolutionary enterprise that was reshaping European thought and culture. The transformation of the discipline had been the work of a succession of scholars as familiar with the accomplishments of the past as they were excited by the independence and aspirations of their own age. De Volder had remarked that among the obstacles to the progress of knowledge were both the devotion to antiquity and the love of novelty,[1] and the development of natural philosophy at Leiden had been marked by a desire on the part of these professors to participate in the innovating temper of their time without repudiating the rich legacy of the past or the traditions of their profession.

The freedom they exercised within the university to weigh the old against the new was seriously curtailed by neither church nor state. To be sure, the curators fended off such intervention by enacting regulations of their own, culminating, indeed, in the eventual dismissal of Heidanus. But the curators do not appear to have sought or intended the actual exclusion of any philosophic alternative, and their regulations pertaining to philosophic deviation were remarkable primarily for their ineffectiveness. In the face of the rising tide of Cartesianism, early efforts to define the limits of acceptable philosophic instruction apparently stemmed from the professors themselves sitting in the academic senate, and those who strove to drive the new philosophy completely from the university were to be found not in the administration but among the faculty. Academic persecution at Leiden was largely the work of the professors, one against

---

[1] *Oratio de novis et antiquis* (Lugduni Batavorum: Apud Cornelium Boutestein, 1709), p. 2.

CONCLUSION

the other, and two major instances of external intervention or administrative authoritarianism, in 1619 and 1676, were passing exceptions rising from crises of state and resulted in no meaningful limitations on philosophic instruction.

Since prohibitions did not constitute an effective barrier against the adoption of philosophic innovations at Leiden, it becomes all the more apparent that the real obstacle confronting the new science within the university was the continuing grip of traditional patterns of thought upon the individual professors themselves. The old philosophy could not simply be proved in error and discarded. Its most fundamental assumptions rested on a different but not a mistaken perception of nature and could not, strictly speaking, be refuted. The new alternatives embodied in the new science could only be made to appear more compelling and more worthwhile, and, though some innovations inspired by seventeenth-century science probably offered welcome relief from strains felt within the old philosophy, others demanded conceptual readjustments that proved difficult for the most sympathetic advocates of philosophic reappraisal.

Among the earliest and most evident effects of the new science on natural philosophy at Leiden had been the gradual corrosion of the perfection and incorruptibility of the heavens. Although a first step towards a homogeneous universe, this corrosion did not as yet threaten any serious disruption of the cosmos and, by stimulating new speculation on the stars, may well have proved as much a source of pleasure as discomfort. Heliocentrism, however, was more disturbing, and the response to Copernicanism was surprisingly unpredictable at Leiden. Accepted as possible by Burgersdijck, it was abruptly rejected by Kyper, who was otherwise relatively imaginative and uninhibited in his thoughts about the heavens. Even after Descartes had integrated both heliocentrism and the mutability of the heavens into a comprehensive new image of the universe, Heereboord, the outspoken champion of freedom and dissent in philosophy, declined to consider the question of heliocentrism, and Senguerdius, like Kyper before him, adopted the geocentrism of Tycho Brahe and appealed to the Bible for support. As the eighteenth century opened and the Newtonian influence was already beginning to make itself felt at Leiden, Senguerdius was teaching a constellation of ideas – absolute motion and a geocentric and finite world – that still called to mind the rejected cosmos.

The erratic career of heliocentrism at Leiden may perhaps betray a deep emotional bias that made the abandonment of geocentrism a difficult and, at first, uncertain step. On the other hand, that erratic history also suggests that the opposition between the ideas of a sun-centered and an earth-centered planetary system was not, after all, among the more decisive or profound aspects of the change in natural philosophy that accompanied the rise of modern science. Other more fundamental concepts in natural philosophy – substance, motion, the very nature and practice of science – revealed at Leiden a more regular and gradual transition that seems to bespeak a deeper, less idiosyncratic philosophic journey in progress.

Nor do all the stages of this journey appear to have been determined only by the confrontation with the new science, for there were tensions within late scholasticism which themselves seemed to call for the kind of relief that the mechanical philosophy momentarily provided. Heereboord had accepted the Cartesian definition of neither matter nor motion, but he found the Cartesian imagery an agreeable alternative to the scholastic principles which now so exasperated him. "No subject in all philosophy is more difficult than that of forms," he declared,[2] and the late scholastics themselves acknowledged the difficulty of understanding first matter. By contrast, the particles of extension in motion were easily and vividly conjured up by the imagination, and their identification with persisting sensible attributes answered to what appears to have been a longing even among the late scholastics for principles with a reality more suggestive of physical existence. When Senguerdius made a weak attempt to reassert form as a principle of physics years after Heereboord, form itself had become the product of those same particles of extension in motion,[3] and 's Gravesande and Musschenbroek were unable to conceive of anything besides that corpuscular imagery that might render natural processes intelligible. That imagery had provided an interval of seemingly clear and simple understanding between the abstruseness of the scholastic principles and the incomprehensibility accepted in Newtonian physics.

While the corpuscular imagery of the mechanical philosophy was accessible but of only passing relevance to the new science, the new perception of motion was essential but elusive. It was difficult to deny what appeared so obvious, that every motion was going somewhere, that natural objects, when they moved, moved toward some proper end or place. With what relentless insistence every stone sought the center of the earth, and with what flawless consistency every planted seed reproduced its parent plant. To perceive motion and change as aimless, oblivious to the nature of bodies and the plan of the cosmos alike, displaying nothing

[2] *Meletemata philosophica* [1654], p. 125.
[3] *Philosophia naturalis,* pp. 7-8 and 27-8.

but brute perseverance, was contrary to common sense and experience. It was still more difficult to acquiesce in a conception of motion that tended to obscure its very existence. "Whoever doubts that there is motion," wrote Burgersdijck, "withdraws all certitude from the senses and at the same time destroys the whole of natural philosophy," [4] but the doctrine of the relativity of motion seemed to shroud its reality in uncertainty. It was a doctrine that repelled no less a mind than Newton's, and he and his eighteenth-century followers continued to insist on an absolute motion which they could not detect.

Nor did the banishment of teleological motion and change from physics prevent a continuing insistence on a deeper teleology in nature as well. Along with physics, metaphysics, ethics and logic, Musschenbroek included among the branches of philosophy a science of teleology, which "investigates the ends, so far as human sagacity can search them out, by reason of which all things in the universe come to pass and all actions, mutations, and motions of things occur." [5] All things, he maintained, had their own particular ends, even if often beyond human understanding, and all things at the same time served the universe as a whole. 's Gravesande had revealed his own similar belief in a purposeful direction guiding nature by undertaking a "Mathematical Demonstration of the Care God Takes to Direct that which Takes Place in this World, Drawn from the Number of Boys and Girls Born Daily." [6]

For Musschenbroek, as well, the ends and purposeful interconnections of all natural things derived from God, who was no less in evidence in the physics of Musschenbroek than in the physics of Burgersdijck a century before. Though 's Gravesande felt obliged in his inaugural address to defend the "mathematicians" against suspicions of irreligion and atheism,[7] the new science and related philosophic systems had in fact betrayed no incompatibility with piety at Leiden. The violent emotionalism that distinguished the clash over Cartesianism from the peaceful introduction of Newtonianism, it is true, had stemmed largely from the theological paranoia – and the rougher university life – that characterized the seventeenth century. But what had been a challenge to a theological system was not inherently a challenge to piety. The modern reader cannot but be struck by the religious feeling that permeates the treatment of the critical subject of motion in de Raey's *Clavis philosophiae naturalis*. Heereboord had maintained that all philosophy, physics being no ex-

---

[4] *Collegium physicum,* p. 63.
[5] *Introductio,* pp. 2-3.
[6] *Oeuvres philosophiques et mathématiques,* Vol. II, p. 221 ff.
[7] "Discours sur l'utilité des mathématiques," pp. 311-2 and 316-7.

ception, was "natural knowledge of God,"[8] and 's Gravesande continued to urge upon his students that the universe testified to a supreme omnipotent intelligence.[9] At the end of his life, when the Enlightenment was in full stride, Musschenbroek elaborated in the *Introductio ad philosophiam naturalem* on how physics revealed the existence, the power, the wisdom and the goodness of God.[10] The new science did not appear at Leiden in the secularizing role which is ascribed to it in European history, and, though the confrontations between the old and the new at Leiden were most bitter when the orthodox theology was threatened, these confrontations were not clashes between the forces of piety and secularism.

Whether contributing to a new piety or shaking confidence in the old, the new science was engendering a growing awareness of an infinitely complex but nonetheless rigorous order that ruled everywhere in nature. Burgersdijck had spoken of the "incredible delight" of investigating natural things,[11] and Heereboord and de Raey called for a return to such investigation with eloquent urgency. But not until de Volder and Senguerdius and the introduction of experimental physics was an active probing of nature actually incorporated into natural philosophy at Leiden. Shortly thereafter, or perhaps simultaneously, came efforts to illustrate some of nature's discovered mathematical regularities as well, and by the early eighteenth century the discipline of physics at Leiden had become identified with the scientific procedures that were constantly revealing new consistencies in nature.

Throughout much of the seventeenth century, however, the new philosophy of Descartes had overshadowed the new science itself at Leiden, and that which inspired the party of "moderns" at the university had not been the promise of what we now know as modern science but the promise of the final fulfillment of traditional philosophic aspirations. By virtue of first principles that were assumed to be not only true but completely comprehensible, Cartesianism seemed at last to offer a clear and comprehensive understanding of nature, an understanding that rested on a complete knowledge of nature's most fundamental natural causes. Philosophers were on the brink of a physics that would be stable and no longer open to dispute. But the professors at Leiden who increasingly identified their physics with the procedures of the new science turned away not only from the hope of such understanding but from the possi-

[8] *Philosophia naturalis cum novis commentariis explicata*, pp. 1-2.
[9] *Philosophiae Newtonianae institutiones*, p. 285.
[10] *Introductio*, pp. 22-3.
[11] *Collegium physicum*, p. 11.

bility of stability and permanence in philosophic knowledge as well. They joined the science of nature to an irrepressible activity that was perpetually questioning old conclusions, perpetually discovering novelty. Something new was discovered in physics everyday, wrote Musschenbroek, and he found good reason to believe that in another century things would appear quite differently in physics than they had in his own day.[12]

The shallowness of the philosophic understanding provided by the new science was to be increasingly obscured by the excitement of this unending discovery and by the consciousness of the greater control of nature that was being won. This is not to say, however, that the science of physics had also been changing from an essentially intellectual to an essentially utilitarian effort. In his last and major text, Musschenbroek listed numerous examples to demonstrate the "enormous" utility of physics,[13] but Burgersdijck had also referred to its "infinite uses," [14] and Musschenbroek's text remained an introduction to "natural philosophy," the philosophy of nature, not the study of its utilization. For many of us today, the image of science has been altered by the awesome achievements of nineteenth and twentieth-century technology, but for the educated Europeans of the seventeenth and eighteenth centuries, if we may consider the history of one of their leading universities as indicative, the issues involved in the birth of modern science pertained essentially to such questions as the nature of motion, space, substance, causality, and knowledge itself.

When the eighteenth century opened, the new science was in fact on its way to becoming the most prestigious among the branches of philosophy, but it was, to be sure, a wayward branch, preoccupied more with endless discovery than with achieving the absolute body of knowledge for which philosophers had yearned. Basic philosophical assumptions of the new science were accepted as obscure or, like relative motion, something short of the actual truth, and the expanding mathematical operations now employed defied efforts to understand them through physical counterparts.[15] The validity of knowledge in natural philosophy was increasingly determined not by its assurance of permanent truth or its compatibility with predetermined principles, but by its applicability and fruitfulness in the continuing investigation of natural phenomena. This was, to be sure, "utility," but utility not for the sake of physical well-being and

---

[12] *Essai de physique*, "Preface."
[13] *Introductio*, p. 16 ff.
[14] *Collegium physicum*, p. 11.
[15] See E. W. Strong, "Newton's 'Mathematical Way,'" *Journal of the History of Ideas*, Vol. XII (1951), pp. 90-110.

convenience, but for the sake of the continuing expansion and reshaping of natural philosophy itself.

In the course of the seventeenth century, there had indeed emerged another conception of "science," a conception that, in its intense commitment to ceaseless expansion and discovery, in its willingness to subordinate philosophic understanding to this commitment, and in its exclusion of all that could be neither measured nor manipulated, became a new ideal prototype in European thought. For all of its subsequent importance in the intellectual life of Europe, however, it only slowly and belatedly gained a place in the curricula of the universities. Unlike the science of the Middle Ages, and despite its own early roots in academic communities, the new science took shape largely outside the schools, and increasingly so during the course of the seventeenth century.[16] Indeed, the history of the discipline of physics at Leiden suggests that, as essentially teaching institutions, even the most progressive universities harbored their own unique dynamic of philosophic conservatism.

The reshaping of physics at Leiden had been obstructed not only by the tenacity of traditional philosophic views, but by the needs and traditional practices of pedagogy as well, and by an inherited tendency to view knowledge as, first and foremost, something to be taught. Physics existed in three states, Kyper had written: the first was a state of "infusion," such as when God granted knowledge to Adam and Solomon without effort on their part, and the second, a state of "invention" wherein unknown theorems were brought to light, false theorems detected, obscure ones clarified, and disordered ones methodically arranged.[17] The third was the state of "discipline," in which physical knowledge was taught by a teacher or learned from books. It was this last state that long dominated the conception of science and knowledge at seventeenth-century Leiden. To be sure, Heereboord objected to Regius' having classified physics not as *scientia* but *doctrina*, which to Heereboord meant everything that could be and was customarily taught.[18] But it was not the stress on teaching to which Heereboord objected; on the contrary, he demurred because the meaning of *doctrina*, "which is common to everything of an intellectual

---

[16] Marie Boas, *The Scientific Renaissance, 1450-1630* (New York: Harper and Brothers, c. 1962), pp. 239-40. A. Rupert Hall, "The Scholar and the Craftsman in the Scientific Revolution," in *Critical Problems in the History of Science*, ed. Marshall Clagett (Madison: The University of Wisconsin Press, 1959), pp. 5-10. A. C. Crombie, *Medieval and Early Modern Science* ([Rev. 2nd ed.]; Garden City, N. Y.: Doubleday and Company, Inc., 1959), Vol. II, pp. 114-5.
[17] *Institutiones physicae*, Vol. I, pp. 10-1.
[18] *Philosophia naturalis cum novis commentariis explicata*, p. 5-6.
[19] ...*quod commune est omnibus habitibus intellectualibus*.... (*Ibid.*)

nature," ¹⁹ was too inclusive, too all-embracing. In contrast to the ancients, complained de Raey, "we Christians devote ourselves wholly to philosophy with hardly any other purpose than to be able to teach it to others." ²⁰

As a discipline, knowledge needed systematization, systematization that lent itself to teaching. Heereboord and Kyper agreed that the "method" whereby physics was to be taught – easily confused at the beginning of the century with the "method" of philosophy itself – was the "synthetic" method, progressing, that is, from the most simple principles of physical things to physical bodies themselves.²¹ It was a method to which the professors of natural philosophy at Leiden attempted to conform throughout the seventeenth and eighteenth centuries, and it was a method that further emphasized the necessity of a system founded on some kind of first principles in the material to be taught. The beginnings of the new science, however, had been disparate and piecemeal, and, prior to Descartes, they were neither correlated in a consistent system of the world nor united by any philosophic first principles. The first achievements of the new science could undermine the traditional system of the discipline of physics, but could not as yet replace it, and while the independent scientist, the investigator and discoverer, might regard the need for such a systematic discipline with less concern, "science" to the professors was well-nigh meaningless without it.

The early impact of the new science, offering no coherent system of its own, was the gradual disintegration of the internal structure of a systematization that continued nonetheless in use. The pedagogical convenience of the traditional system remained, while the conviction in its philosophic soundness began to wane.²² Burgersdijck, foremost of the Dutch Aristotelians, could entertain the possible invalidity of the whole of his cosmic physics by acknowledging that Copernicus might have been right. Heereboord drifted anchorless in a sea of opinions and speculations that had been loosed by the loss of confidence in peripatetic thought. His was the plight of an eclectic, whose thought de Volder characterized as utter confusion, acknowledging contradictory principles or none at all.²³ Notorious for his tirades against the yoke of tradition, Heereboord nonetheless found no acceptable alternative to the scholastic system for the discipline of physics.

---

²⁰ *Clavis philosophiae naturalis,* p. 4.
²¹ *Philosophia naturalis cum novis commentariis explicata,* pp. 6-7. *Institutiones physicae,* Vol. I, pp. 17-8.
²² See Reif, "Natural Philosophy," p. 323.
²³ *Disputationes philosophicae de rerum naturalium principiis,* p. 7.

This is by no means to suggest that pedagogical needs were the major source of the concern for system in philosophy (though the question of the influence of pedagogical prototypes on European philosophic thought and writing is perhaps worth raising). The apparent order in nature as well as the coherence and consistency that "truth" itself would seem necessarily to entail might appear to make system requisite in any philosophy that seriously claimed to be true; de Volder found the eclectic's jumbled philosophy incompatible with the beauty of the universe, "in which all things are perceived not dispersed and in disunion but joined and gathered together in a certain and admirable order." [24] But the pedagogical need for system was one of immediate application and could not be held in abeyance while one worked out a new system of the universe more compatible with recent discoveries and new insights. Where a new system was not available, the demands of the lecture hall called for the continued use of the old, and the retention of the old system generally meant the retention of old facts and concepts as well. The priorities of philosophy and pedagogy were not always harmonious.

It was Descartes, his *Principia philosophiae* having indeed been written as a textbook,[25] who provided those inclined to the newer trends in European thought with a comprehensive system of natural philosophy. How much this had meant to seventeenth-century pedagogues of the "modern" bent is perhaps suggested by le Clerc's recollection that among the things that were troubling de Volder in the years immediately preceding his retirement was the prospect of having to find an alternative to the Cartesian system.[26] Beyond Descartes, the new science had still been provided with no clear and teachable system, and it was no little contribution to his own influence as well as Newton's when 's Gravesande succeeded in drawing such a system from Newtonian science.

In the physics taught by 's Gravesande, however, the philosophic system was no longer alone in establishing the scope and presentation of the discipline. Already in the physics of Senguerdius, whose sense of philosophic system and coherence was not the keenest, experimentation had also played a decisive role in determining the organization of courses, material to be taught, and the manner of instruction. The discipline of physics was now distinguished as much by its unique method, a pedagogical adaptation of the experimental procedures of the new science, as by its unique body of organized knowledge. Even when joined to a more

[24] *Ibid.*
[25] Mouy, *Le Développement de la physique cartésienne*, p. 18. Koyré, *Newtonian Studies*, p. 74.
[26] See above, pp. 111-2.

CONCLUSION

sophisticated rational system by 's Gravesande and Musschenbroek, this unique method retained its central importance in the discipline, which acquired thereby not only a new kind of internal coherence but a new kind of teaching that was more compatible with the new science as well.

The traditional procedures of teaching were not conducive to the new science and had obstructed its further advance in universities. At late seventeenth-century Leiden, printed disputations had begun to appear with geometric diagrams, algebraic formulations, and detailed explanatory illustrations of experimental instruments.[27] The effort to imagine the handling of this material in actual oral disputation suggests how unsuitable to the new science the traditional pedagogy was. Nor could lectures alone adequately teach a science founded on calculation and the manipulation of nature.

On the other hand, it is likely that the pedagogical methods of 's Gravesande and Musschenbroek appeared unattractive to many academics who continued to cling to traditional images of the professorial profession. Musschenbroek had referred to those who still considered experiments little more than juvenile distractions, and there were surely many who, despite the popular appeal of such distractions,[28] still found little gratification in the idea of their tinkering with machines rather than lecturing from the podium or presiding over disputations. Moreover, the pedagogue had reason to place high value on the stability of his discipline. For the late scholastics, their pedagogical or disciplinary system was a highly-prized achievement, and to be faced with its perpetual undoing by continual change and novelty within the subject matter was not a reassuring prospect. To have identified a philosophic discipline with a method that was constantly producing novelty was hardly to render it more appealing to many professional educators.

Even where the philosophical and pedagogical obstacles were overcome, however, the universities could not hope to play a prominent role in the active pursuits of modern science until the status of the discipline of physics within the university had also changed. Medieval science had

---

[27] See Schuyl, *De vi corporum elastica*; Johannes Kopeczi, *Disputatio philosophica de cometis, secunda bipertita,* sub praesidio D. Johannis de Raei (Lugduni Batavorum: Apud Viduam et Haeredes Johannis Elsevirii, 1666); Joannes Fransiscus de Witte van Schooten, *Dissertatio philosophica inauguralis de solido, ejusque partium, nec non hemisphaeriorum concavorum, et cylindrorum solidorum cohaerentia, pro gradu* (Lugduni Batavorum: Apud Viduam Cornelii Boutesteyn, 1712).

[28] See J. B. Morrell's brief reference to the student appeal of experimental demonstrations in the lecture halls of late eighteenth-century Edinburgh; "The University of Edinburgh in the Late Eighteenth Century: Its Scientific Eminence and Academic Structure," *Isis*, Vol. LXII (1971), p. 161.

developed largely within the framework of the philosophy courses offered at the universities,[29] but the new science had rapidly become too complex for the philosophy curriculum, whose introductory character had been rendered all the more elementary by educational reformers of the sixteenth and early seventeenth century.[30] The philosophy curriculum as a whole continued to be organized as primarily a preparation for the higher faculties through the eighteenth century, and consequently, though others may have protested with de Raey against philosophy's being but "the handmaid and contemptible slave of other studies," [31] physics remained within the framework of a lower faculty where few still lingered to seek a degree.[32] At eighteenth-century Leiden, the study of physics indeed acquired an unprecedented influence, partly as a result of the growing prestige of the natural sciences throughout Europe in general and partly as a result of the personal abilities of 's Gravesande and Musschenbroek. Nonetheless, even Leiden, confronted with no rival scientific institutions of the stature of the scientific societies in London and Paris, could not assume the function of these societies, to which Musschenbroek gave much of the credit for having continued the scientific progress that had begun in the seventeenth century.[33] Leiden had been among the first and most persistent in incorporating experimental physics within the traditional academic curriculum, but even at Leiden research facilities for

---

[29] Crombie, *Medieval and Early Modern Science*, Vol. II, pp. 114-5. Boas, *The Scientific Renaissance*, p. 239.
[30] Ong, *Ramus*, pp. 136-7.
[31] *Clavis philosophiae naturalis*, p. 3.
[32] Kuenen, *Het Aandeel van Nederland*, p. 30.
[33] *Introductio*, p. 4.

The faculty at Leiden did not relish the idea of anyone else assuming this function, however. The oldest of the Dutch scientific societies was that formally established at Haarlem in 1752 (Kuenen, *op. cit.*, p. 7; *Bronnen*, Vol. V, p. 205*), and the academic senate at Leiden, with its own interpretation of the causes of the decline of the French and British universities, rather ungraciously opposed the efforts of the Haarlem society to acquire the official support of the States General in 1759 (*Bronnen*, Vol. V, pp. 204*-6* and 208*-10*). In 1760, the senate informed the university curators of its views (*Ibid.*, p. 209*):
"Although the Haarlem Society, in so far as nothing pertaining to the sciences is taught there orally, nor doctoral degrees granted, etc., is of a different nature [than the university], experience has nonetheless shown how much the glory of more than one university has been beclouded and darkened when public authority was also bestowed upon other societies and when illustrious persons consented to become members of the same."
"This has been the experience in England, for the universities of Cambridge and Oxford have lost much of their previous brilliance and luster since the Royal Society in London was established by public authority. The University of Paris, so famous in former times, is scarcely mentioned anymore since the Royal Academy of Sciences there was brought to bloom by the personal protection of the king and the accession of illustrious persons as honorary members. ..."

the students themselves had to wait until the nineteenth century,[34] and facilities for that "concerted labor of many philosophers" which, for Musschenbroek, characterized the scientific societies [35] was as yet scarcely thought of in a university context. The first beginnings of a new orientation towards research in the universities were only just stirring in the eighteenth-century revival of the universities in Germany.

It has been asserted that the founding of the University of Halle by the Elector of Brandenburg in 1694 and of the University of Göttingen by the Hanoverian dynasty in 1737 marked a major turning point in the history of the European universities, the establishment of the first "modern" universities.[36] The University of Halle embraced a new practical concern with political and social usefulness, to which Göttingen added a new emphasis on research and individual scholarship.[37] Both schools welcomed the new science [38] and defended the principle of freedom in philosophical teaching and research, preparing the way, at least at Göttingen, for the gradual rise of the faculty of philosophy from a position of subordination and neglect to one of supremacy within the university.[39] The eighteenth century also witnessed the flowering of the University of Edinburgh, whose philosophy and medical faculties and their teaching of the natural sciences made Edinburgh one of the most prestigious centers of higher learning in Europe.[40]

The rise of Edinburgh in particular was felt at Leiden in the gradual but unrelieved decline in the number of students that began after the first

---

[34] Vollgraff, "Leidsche Hoogleeraren," pp. 189-90.
Customarily, the students interested in natural science went on for a degree in the higher faculty of medicine (Kuenen, *op. cit.*, p. 30), where active original research might indeed be undertaken by the students, an outstanding case in point being the vivisections and experiments carried out by Jan Swammerdam for his doctoral disputation at Leiden in 1667 (*Nieuw Nederlandsch Biografisch Woordenboek*, Vol. X, 996).

[35] *Introductio*, p. 4.

[36] Paulsen, *Geschichte des gelehrten Unterrichts*, Vol. I, p. 263. D'Irsay, *Histoire des universités*, Vol. II, p. 91. Beck, *Early German Philosophy*, pp. 306-7.

[37] Paulsen, *op. cit.*, Vol. I, pp. 263 and 550. Beck, *op. cit.*, pp. 158-9 and 307. D'Irsay, *loc. cit.* W. H. Bruford, *Germany in the Eighteenth Century: The Social Background of the Literary Revival* (Cambridge: At the University Press, 1965), pp. 237, 241 and 243-5.

[38] Paulsen, *op. cit.*, Vol. I, pp. 536, 540 and 548-9. Bruford, *op. cit.*, p. 245.

[39] Beck, *Early German Philosophy*, pp. 290 and 306-7. Paulsen, *op. cit.*, Vol. I, p. 263. D'Irsay, *Histoire des universités*, Vol. II, pp. 94-5. Bruford, *op. cit.*, pp. 236 and 239-40.

[40] D. B. Horn, *A Short History of the University of Edinburgh 1556-1889* (Edinburgh: The University Press, c. 1967), pp. 43, 47-8, 52-3 and 56. F. Sherwood Taylor, "The Teaching of the Physical Sciences at the End of the Eighteenth Century," in *Natural Philosophy Through the 18th Century*, ed. Ferguson, p. 155.

quarter of the century,[41] but both Edinburgh and, to a lesser degree, the German universities owed much to Leiden, particularly in the realm of medicine. Nearly the whole of Edinburgh's newly created medical faculty in the early eighteenth century had studied at the Dutch university,[42] and it was particularly von Haller, "the presiding genius of eighteenth-century medicine," who brought celebrity to the medical faculty at Göttingen.[43] Nor, despite the rise of such vigorous rivals, did Leiden soon relinquish her European preeminence. She continued to attract students from both Britain and Germany,[44] and such notorious French *philosophes* as La Mettrie and d'Holbach had sat among them as well. If we may believe Diderot's *Encyclopédie*, Leiden was in 1765 still to be considered the first university of Europe.[45] Due particularly to the work of Bernhard Siegfried Albinus, who taught at Leiden from 1719 to 1770, the university remained a major center of anatomical studies, while Hieronymus David Gaubius was, until his death in 1780, among the most influential teachers of pathology.[46] 's Gravesande and Musschenbroek – to say nothing of Boerhaave, who died only in 1738 – likewise continued to draw students from all over Europe,[47] and their immediate successors, Allamand and Lulofs, also enjoyed a reputation abroad, Allamand as a disciple of 's Gravesande in experimental physics and Lulofs as an able astronomer.[48]

[41] de Vrankrijker, *Vier Eeuwen*, p. 44. Horn, *op. cit.*, p. 67.
[42] Horn, *op. cit.*, pp. 42-3. D'Irsay, *Histoire des universités*, Vol. II, pp. 82-3. E. D. Baumann, *Uit drie Eeuwen Nederlandse Geneeskunde* (Amsterdam: H. Meulenhoff, [1951]), pp. 220-1.
[43] Richard H. Shryock, *The Development of Modern Medicine, An Interpretation of the Social and Scientific Factors Involved* (Philadelphia: University of Pennsylvania Press, 1936), p. 62. D'Irsay, *op. cit.*, p. 98. Baumann, *op. cit.*, pp. 219 and 225.
[44] Hans, *New Trends in Education*, p. 24. Bruford, *Germany in the Eighteenth Century*, p. 258.
[45] Diderot and d'Alembert, eds., *Encyclopédie, ou dictionnaire raisonné des sciences, des arts et des métiers* (Paris: Briasson, 1751-65), Vol. IX, p. 451.
[46] Erik Nordenskiöld, *The History of Biology*, trans. Leonard Bucknall Eyre (New York and London: Alfred A. Knopf, 1928), p. 258. Castiglioni, *A History of Medicine*, p. 608. *Nieuw Nederlandsch Biografisch Woordenboek*, Vol. III, 431-2; Vol. IV, 22-4.
[47] Voltaire, *The Complete Works*, ed. Theodore Besterman, *et al.* (Geneve: Institut et Musée Voltaire; Toronto and Buffalo: University of Toronto Press; 1968-   ), Vol. 88, p. 180. Condorcet, *Oeuvres*, ed. A. Condorcet O'connor and M. F. Arago (Paris: Firmin Didot Frères, 1847-9), Vol. II, pp. 126-7.
[48] Brunet, *Les Physiciens hollandais*, p. 148.
Lulofs, who had succeeded 's Gravesande in mathematics and astronomy and died in 1768, was an accomplished astronomer who attempted, with no little frustration, to use the university observatory for his own scientific observations as well as for teaching and public demonstrations (*Bronnen*, Vol. V, p. 237; Kuenen, *Het Aandeel van Nederland*, pp. 14-5; de Sitter, *Short History of the Observatory*, pp. 17-8).

CONCLUSION

But Allamand and Lulofs were not, after all, of the same caliber as Musschenbroek and 's Gravesande, and by Allamand's death in 1787 – Lulofs had died in 1768 – the practice and prestige of the natural sciences were no longer what they had been at Leiden a quarter of a century before.[49] Despite revived but ultimately fruitless aspirations in the very decade of Allamand's death,[50] a continuing deterioration of the natural sciences at Leiden contributed now to the general decline of the university, one more symptom, it appeared, of the failing energies of the Dutch Republic. The political power of the republic had been permanently undermined during the long wars against Louis XIV, and her economic supports had gradually but inexorably crumbled during the course of the following century, providing a somber setting for the university's fading image in the last decades. In 1784, in Leiden itself, the visiting American merchant Elkanah Watson was struck by the late evening hush of a city that was now too large for its diminishing population, and though his passing comment confirmed the continuing fame of the university, whose professors "have been the most eminent in Europe," the emphasis was on the past.[51]

A remarkable past it had been, and it owed much as well to the achievements of many who individually had been something less than the most eminent in Europe. Through the generations of Leiden's faculty there had continued a line of philosophy professors who seldom could aspire to even a second rank among the scientists and philosophers of Europe and yet who distinguished their university by their sympathetic and

---

Allamand had been added to the faculty as professor of mathematics and philosophy in 1749 in response to the increased interest in the natural sciences at the university (Siegenbeek, *Geschiedenis der Leidsche Hoogeschool*, Vol. I, p. 275). Because of the excellence of the instruction offered by Lulofs and Musschenbroek, Allamand himself turned to the teaching of natural history, but, being a disciple of 's Gravesande, he also began offering lessons in experimental physics and assumed the responsibility for the university's offerings in experimental physics following Musschenbroek's death (*Nieuw Nederlandsch Biografisch Woordenboek*, Vol. I, 75-7; *Bronnen*, Vol. V, pp. 462 and 211*-2*). He had participated in Musschenbroek's experiments in electricity even before joining the faculty and possessed his own large collection of instruments, which he left to the university.

[49] Vollgraff, "Leidsche Hoogleeraren," p. 189. Sassen, *Geschiedenis van de Wijsbegeerte in Nederland*, p. 234.

[50] Woltjer, *De Leidse Universiteit*, pp. 60-1.

[51] [Elkanah Watson], *A Tour of Holland in MDCCLXXXIV* (Worcester, Mass.: Isaiah Thomas, 1790), pp. 105 and 99.

By the end of the eighteenth century, Leiden had indeed dropped from a population of about 60,000 a century before to approximately a little more than half that size, a decline in population shared by most of the other towns of Holland as well (Faber, "Population Changes," pp. 56-7; Mols, *Introduction*, Vol. II, p. 253).

continuing interest in the radical philosophic innovations of their time. Short on tradition and governed by the representatives of a broad-minded urban oligarchy, the University of Leiden, with but few and momentary exceptions, had shielded and encouraged the efforts of these professors to find their way through the contending claims of the old philosophy and the new science. As a consequence, Leiden had emerged in the seventeenth and eighteenth centuries as the European university within which the import of the beginnings of modern science had been most fully recognized and the most persistent and successful effort made to incorporate those beginnings within the traditional curriculum of higher learning.

SELECTED BIBLIOGRAPHY

SECONDARY SOURCES

Pertaining to the University of Leiden:

van Arkel, D. "Leids Studentenleven in de 16e, 17e en 18e Eeuw," *Geschiedboek van het Leidsche Studenten Corps.* Leiden: H. E. Stenfert Kroese, 1950.

Baumann, E. D. *Uit drie Eeuwen Nederlandse Geneeskunde.* Amsterdam: H. Meulenhoff, [1951].

Bierens de Haan, David. *Bibliographie néerlandaise historique-scientifique des ouvrages importants dont les auteurs sont nés aux 16e, 17e et 18e siècles, sur les sciences mathématiques et physiques avec leurs applications.* Rome: Imprimerie des sciences mathématiques et physiques, 1883 (Extract from the *Bullettino di bibliografia di storia delle scienze matematiche e fisiche,* Vols. XIV-XVI, 1881-3.)

Bohatec, Josef. *Die cartesianische Scholastik in der Philosophie und reformierten Dogmatik des 17. Jahrhunderts.* Leipzig: A. Deichert, 1912.

Brunet, Pierre. *Les Physiciens hollandais et la méthode expérimentale en France au XVIIIe siècle.* Paris: Albert Blanchard, 1926.

Cohen, Gustave. *Écrivains Français en Hollande dans la première moitié du XVIIe siècle.* La Haye: Martinus Nijhoff; Paris: Édouard Champion; 1921.

Costabel, Pierre. "'s Gravesande et les forces vives ou des vicissitudes d'une expérience soi-disant cruciale," *Mélanges Alexandre Koyré,* Vol. I: *L'Aventure de la science.* Paris: Hermann, c. 1964.

Cramer, Jan Anthony. *Abraham Heidanus en zijn Cartesianisme.* Utrecht: J. van Druten, 1889.

Crommelin, Claude August. *Instrumentmakerskunst en proefondervindelijke Natuurkunde.* Leiden: Eduard Ijdo, 1925.

—. "Physics and the Art of Instrument Making at Leyden in the 17th and 18th Centuries," *Lectures on Physics and Physiology Delivered in the University of Leyden During the Second Netherlands Week for American Students July 5-10, 1926.* Leyden: A. W. Sijthoff, 1926.

Dibon, Paul. "L'Influence de Ramus aux universités Néerlandaises du 17e siècle," *Actes du XIème Congrès International de Philosophie,* Volume XIV (1953), pp. 307-11.

—. *La Philosophie néerlandaise au siècle d'or,* Vol. I: *L'Enseignement philosophique dans les universités à l'époque pré-cartésienne, 1575-1650.* Paris, etc.: Elsevier Publishing Company, 1954. The authoritative study of philosophy in the Dutch universities during the period covered. The second volume has yet to appear.

Dijksterhuis, E. J., et al. *Descartes et le cartésianisme hollandais.* Paris: Presses Universitaires de France; Amsterdam: Editions Françaises d'Amsterdam, 1950. Contains valuable articles by Paul Dibon and C. Louise Thijssen-Schoute.

Durnin, Richard G. "The University of Leyden, and America," *Paedagogica Historica,* Vol. VIII (1968), pp. 335-50.

Eekhof, A. *De Theologische Faculteit te Leiden in de 17de Eeuw.* Utrecht: G. J. A. Ruys, 1921.

Hans, Nicholas. "Holland in the Eighteenth Century – *Verlichting* (Enlightenment)," *Paedagogica Historica,* Vol. V (1965), pp. 14-37.

Israëls, A. H. "De Verdiensten der Nederlanders in het Verspreiden en Uitbreiden der Harveyaansche Ontdekking," *Nederlandsch Tijdschrift voor Geneeskunde* (tevens orgaan der Nederlandsche Maatschappij tot Bevordering der Geneeskunst), Vol. IV (1860), pp. 361-73.

Jorissen, W. P. *Het chemisch (thans anorganisch chemisch) Laboratorium der Universiteit te Leiden van 1859-1909 en de chemische Laboratoria dier Universiteit vóór dat Tijdvak en hen, die er in doceerden.* Leiden: A. W. Sijthoff, 1909.

Kist, N. C. *Bijdragen tot de vroegste Geschiedenis en den Toekomstigen Bloei der Hoogeschool te Leiden.* Leiden: P. H. van den Heuvell, 1850. Brief, but includes documents.

Kroon, Just Emile. *Bijdragen tot de Geschiedenis van het geneeskundig Onderwijs aan de Leidsche Universiteit 1575-1625.* Leiden: S. C. van Doesburgh, 1911.

Kuenen, J. P. *Het Aandeel van Nederland in de Ontwikkeling der Natuurkunde gedurende de laatste 150 Jaren.* Bataafsch Genootschap der Proefondervindelijke Wijsbegeerte te Rotterdam, 1919. Provides an informative first chapter on scientific activity in the Netherlands in the eighteenth century.

Land, J. P. N. "Arnold Geulincx and His Works," *Mind,* Vol. XVI, pp. 223-42.

—. "Arnold Geulincx te Leiden (1658-1669)," *Mededeelingen, Kon. Akademie van Wetenschappen, Amsterdam, Afdeeling Letterkunde,* series 3, Vol. III (1887), pp. 277-327.

—. *Arnold Geulincx und seine Philosophie.* Haag: Martinus Nijhoff, 1895.

—. "Schotsche Wijsgeeren aan Nederlandsche Hoogescholen," *Verslagen en Mededeelingen, Kon. Akademie van Wetenschappen, Amsterdam, Afdeeling Letterkunde,* series 2, Vol. VII (1878), pp. 168-84.

—. "Philosophy in the Dutch Universities," *Mind,* old series, Vol. III (1878), pp. 87-104.
Lindeboom, G. A., ed. *Boerhaave and His Time.* Leiden: E. J. Brill, 1970.
Russell, L. J. "The Correspondence Between Leibniz and de Volder," *Proceedings of the Aristotelian Society,* new series, Vol. XXVIII (1927-8), pp. 155-76.
Sassen, Ferdinand. "Adriaan Heereboord (1614-1661), De Opkomst van het Cartesianisme te Leiden" *Algemeen Nederlands Tijdschrift voor Wijsbegeerte en Psychologie,* Vol. XXXVI (1942-3), pp. 12-22.
—. *Geschiedenis van de Wijsbegeerte in Nederland tot het Einde der negentiende Eeuw.* Amsterdam en Brussel: Elsevier, 1959. An excellent introduction to the development of Dutch philosophy.
—. "Het oudste wijsgeerig Onderwijs te Leiden (1575-1619)," *Mededeelingen, Kon. Akademie van Wetenschappen, Amsterdam, Afdeeling Letterkunde,* new series, Vol. IV (1941).
Schneppen, Heinz. *Niederländische Universitäten und Deutsches Geistesleben von der Gründung der Universität Leiden bis ins späte 18. Jahrhundert.* Münster, Westfalen: Aschendorff, 1960.
Schotel, G. D. J. *De Academie te Leiden in de 16ᵉ, 17ᵉ en 18ᵉ Eeuw.* Haarlem: Kruseman en Tjeenk Willink, 1875. Though more compact than the following work, it nonetheless adds some useful and interesting information.
Siegenbeek, Matthijs. *Geschiedenis der Leidsche Hoogeschool, van hare Oprigting in den Jare 1575, tot het Jaar 1825.* 2 vols.; Leiden: S. en J. Luchtmans, 1829-32. Still the most comprehensive history of the University of Leiden during its first centuries.
de Sitter, Willem. *Short History of the Observatory of the University at Leiden, 1633-1933.* Haarlem: Joh. Enschedé en zonen, [1933].
Stearn, William Thomas. "The Influence of Leyden on Botany in the Seventeenth and Eighteenth Centuries," *Early Leyden Botany.* Universitaire Pers Leiden; Assen: Van Gorcum en Comp. N. V., 1961.
Suringar, G. C. B. *Bijdragen tot de Geschiedenis van het geneeskundig Onderwijs aan de Leidsche Hoogeschool, van de Stichting der Universiteit in 1575 tot aan den Dood van Boerhaave, 1738.* Twelve articles offprinted from the *Nederlandsch Tijdschrift voor Geneeskunde,* 1860-1866. These articles offer the most complete study of the medical faculty and medical instruction at Leiden throughout the period.
Thijssen-Schoute, C. Louise. *Nederlands Cartesianisme. Verhandelingen, Kon. Akademie van Wetenschappen, Amsterdam, Afdeling Letterkunde,* new series, Vol. LX (1954). A massive reservoir of information on Cartesianism in the Netherlands. It includes a summary in French.
Vander Haeghen, Victor. *Geulincx: Étude sur sa vie, sa philosophie et ses ouvrages.* Gand: Ad. Hoste, 1886.
Vollgraff, J. A. "Leidsche Hoogleeraren in de Natuurkunde in de 16ᵈᵉ, 17ᵈᵉ en 18ᵈᵉ Eeuw," *Jaarboekje voor Geschiedenis en Oudheidkunde van Leiden*

en Rijnland, orgaan der Vereeniging "Oud-Leiden." Leiden: A. W. Sijthoff, 1913.

—. "Pierre de la Ramée (1515-1572) et Willebrord Snell van Royen (1580-1626)," *Janus,* Vol. XVIII (1913), pp. 595-625.

de Vrankrijker, A. C. J. *Vier Eeuwen Nederlandsch Studentenleven.* Voorburg: Uitgeverij Boot N. V., [1939].

Woltjer, J. J. *De Leidse Universiteit in Verleden en Heden.* Universitaire Pers Leiden, 1965. A very brief but useful introduction to the history of the university.

European Higher Education in the Early Modern Period and the Response to the New Science:

Addy, George M. "Alcalá Before Reform – the Decadence of a Spanish University," *The Hispanic American Historical Review,* Vol. XLVIII (1968), pp. 561-85.

—. *The Enlightenment in the University of Salamanca.* Durham, N. C.: Duke University Press, 1966.

Allen, Phyllis. "Scientific Studies in the English Universities of the 17th Century," *Journal of the History of Ideas,* Vol. X (1949), pp. 219-53.

Ashby, Eric. *Technology and the Academies: An Essay on Universities and the Scientific Revolution.* London: Macmillan and Co., Ltd.; New York: St. Martin's Press, Inc., 1958. Brief and written with an eye on twentieth-century problems; nonetheless, it provides a useful sketch of the response of the universities to modern science over the last four centuries.

Beck, Lewis White. *Early German Philosophy: Kant and His Predecessors.* Cambridge, Mass.: The Belknap Press of Harvard University Press, 1969.

Borgeaud, Charles. *Histoire de l'Université de Genève: L'Académie de Calvin, 1559-1798.* Genève: Georg et Co., 1900.

Boutroux, Pierre. "L'Enseignement de la Mécanique en France au XVIIe siècle," *Isis,* Vol. IV (1921-2), pp. 276-94. Stresses how seventeenth-century textbooks presented a distorted picture of the pattern of scientific change during the century.

Bruford, W. H. *Germany in the Eighteenth Century: The Social Background of the Literary Revival.* Cambridge: At the University Press, 1965.

Charlton, Kenneth. *Education in Renaissance England.* London: Routledge and Kegan Paul; Toronto: University of Toronto Press; 1965.

Clark, George. *The Seventeenth Century.* 2nd ed.; New York; Oxford University Press, 1961.

Compayré, Gabriel. *Histoire critique des doctrines de l'éducation en France depuis le seizième siècle.* 7th ed.; 2 vols.; Paris: Hachette et Cie, 1904.

Costello, W. T. *The Scholastic Curriculum at Early 17th Century Cambridge.* Cambridge, Mass.: Harvard University Press, 1958.

Curtis, Mark H. *Oxford and Cambridge in Transition, 1558-1642.* Oxford: At the Clarendon Press, 1959.

Debus, Allen G. *Science and Education in the Seventeenth Century*: The

Webster-Ward Debate. London: Macdonald; New York: American Elsevier Inc.; 1970. Consists largely of the texts of the seventeenth-century English debate, but also includes a lengthy and revealing introductory study of the origins of university criticism in the sixteenth and early seventeenth century with particular emphasis on the traditions of Renaissance magic and chemistry.

Delannoy, Paul. *L'Université de Louvain.* Paris: Auguste Picard, 1915.

Galama, Sybrand H. M. *Het wijsgerig Onderwijs aan de Hogeschool te Franeker, 1585-1811.* Franeker: T. Wever, 1954.

Grant, Alexander. *The Story of the University of Edinburgh During Its First Three Hundred Years.* 2 vols.; London: Longmans, Green and Co., 1884.

Günther, S. "Die mathematischen und Naturwissenschaften an der nürnbergischen Universität Altdorf," *Mitteilungen, Vereins für Geschichte der Stadt Nürnberg,* No. 3 (1881), pp. 1-36.

Gunther, R. T. *Early Science in Cambridge.* Oxford, 1937.

Hall, A. Rupert. "The Scholar and the Craftsman in the Scientific Revolution," *Critical Problems in the History of Science,* ed. Marshall Clagett. Madison: The University of Wisconsin Press, 1959. An illuminating essay that gives much attention to the relationship between university education and the development of the new science.

Hans, Nicholas. *New Trends in Education in the Eighteenth Century.* London: Routledge and Kegan Paul Ltd., 1951. The focus is on England.

Herr, Richard. *The Eighteenth-Century Revolution in Spain.* Princeton, N. J.: Princeton University Press, 1958.

Horn, D. B. *A Short History of the University of Edinburgh, 1556-1889.* Edinburgh: The University Press, c. 1967.

d'Irsay, Stephen. *Histoire des universités françaises et étrangères des origines à nos jours.* 2 vols.; Paris: Éditions Auguste Picard, 1933-5. The only work I have found that attempts a general history of the European universities through the nineteenth century. Though by no means as comprehensive and detailed as Hastings Rashdall's history of the universities in the Middle Ages, it is a fine survey and provides a valuable sequel to Rashdall's study.

Jones, Richard Foster. *Ancients and Moderns, A Study of the Rise of the Scientific Movement in Seventeenth-Century England.* 2nd ed.; St. Louis: [Washington University], 1961. Considers Puritan criticisms of the universities and the traditional curriculum.

Jourdain, Charles. *Histoire de l'Université de Paris au XVIIe et au XVIIIe siècle.* Paris: L. Hachette et Cie, 1862-6.

Kagan, Richard L. "Universities in Castile, 1500-1700," *Past and Present,* No. 49 (Nov. 1970), pp. 44-71.

Kamen, Henry. "Intellectuals on Trial," *Encounter,* Vol. XXVIII (1967), pp. 10-9. Puts much of the blame for the intellectual repression in the Spanish universities in the late sixteenth century on the academics themselves.

Kearney, Hugh. *Scholars and Gentlemen: Universities and Society in Pre-Industrial Britain, 1500-1700.* London: Faber and Faber, 1970.

Knox, H. M. *Two Hundred and Fifty Years of Scottish Education, 1696-1946*. Edinburgh and London: Oliver and Boyd, 1953.

Loncq, G. J. *Historische Schets der Utrechtsche Hoogeschool tot hare Verheffing in 1815*. Utrecht: J. L. Beijers en J. van Boekhoven, 1886.

Mallet, Charles Edward. *History of the University of Oxford*. 3 vols.; London: Methuen and Co., Ltd., 1924-7.

McLachlan, Herbert. *English Education Under the Test Acts: Being the History of the Non-Conformist Academies, 1662-1820*. Manchester: University Press, 1931.

Morgan, Alexander. *Scottish University Studies*. Oxford University Press; London: Humphrey Milford, 1933.

Morison, Samuel Eliot. *The Founding of Harvard College*. Cambridge, Mass.: Harvard University Press, 1935. Includes many observations about the European university tradition from which Harvard sprang.

Morrell, J. B. "The University of Edinburgh in the Late Eighteenth Century: Its Scientific Eminence and Academic Structure," *Isis, Vol.* LXII (1971), pp. 158-71.

Mousnier, Roland, *Paris au XVIIe siècle*. "Les cours de Sorbonne," Histoire moderne et contemporaine; Paris: Centre de documentation universitaire, [1961]. Includes a very good description of the university and its organization in the seventeenth century.

Mullinger, James Bass. *The University of Cambridge*. 3 vols.; Cambridge: At the University Press, 1873-1911.

Nicolson, Marjorie. "The Early Stages of Cartesianism in England," *Studies in Philology*, Vol. XXVI (1929), pp. 356-74.

Ong, Walter J. *Ramus, Method, and the Decay of Dialogue*. Cambridge, Mass.: Harvard University Press, 1958. Offers enlightening insights into the effect of pedagogical traditions and concerns in sixteenth-century thought.

Ornstein, Martha. *The Rôle of the Scientific Societies in the Seventeenth Century*. 3rd ed.; Chicago: The University of Chicago Press, 1938. Still the most comprehensive attempt to assess the value of the scientific societies and the failings of the universities with regard to the new science of the seventeenth century.

Paulsen, Friedrich. *Geschichte des gelehrten Unterrichts*. 3rd ed.; 2 vols.; Leipzig: Veit, 1919-21. A masterful study of the German universities in the context of German social and intellectual history.

—. *The German Universities and University Study*. Trans. Frank Thilly and William W. Elwang; New York: Charles Scribner's Sons, 1906.

Peacock, George. *Observations on the Statutes of the University of Cambridge*. London: John W. Parker; Cambridge: J. and J. J. Deighton; 1841. A clear depiction, in the context of an appeal for reform, of the scholastic procedures established by statute.

Puschmann, Theodor. *A History of Medical Education*. Trans. and ed. Evan H. Hare; London: H. K. Lewis, 1891.

Randall, John Herman, Jr. *The School of Padua and the Emergence of Modern Science.* Padova: Editrice Antenore, 1961. Stresses the vitality of Aristotelianism in Padua and the framework it provided for the antecedents of Galilean science.

Rashdall, Hastings. *The Universities of Europe in the Middle Ages.* New ed. by F. M. Powicke and A. B. Emden; 3 vols.; Oxford University Press, 1936. The classic work on the first centuries of the European universities.

Reif, Mary Richard. "Natural Philosophy in Some Early Seventeenth Century Scholastic Textbooks." Unpublished Ph.D. dissertation; Saint Louis University, 1962.

— [Reif, Patricia]. "The Textbook Tradition in Natural Philosophy, 1600-1650," *Journal of the History of Ideas,* Vol. XXX (1969), pp. 17-32.

Rybka, Eugeniusz. *Four Hundred Years of the Copernican Heritage.* Jagellonian University Jubilee Publications, Vol. XVIII; Cracow, 1964. Some information, but not much, on the astronomical tradition at the University of Cracow and the Italian universities.

Schrader, Wilhelm. *Geschichte des Friedrichs-Universität zu Halle.* 2 vols.; Berlin: Ferd. Dümmlers Verlagsbuchhandlung, 1894.

Schwickerath, Robert. *Jesuit Education: Its History and Principles Viewed in the Light of Modern Educational Problems.* St. Louis, Mo.: B. Herder, 1903.

Sicard, Augustin. *Les études classiques avant la révolution.* Paris: Perrin et Cie, 1887.

Taton, René, ed. *Enseignement et diffusion des sciences en France au XVIIIe siècle.* Paris: Hermann, c. 1964.

Thompson, Craig R. *Universities in Tudor England.* Washington: The Folger Shakespeare Library, 1959. A brief but useful pamphlet.

Van der Essen, Léon. *Une institution d'enseignement supérieur sous l'ancien régime: L'Université de Louvain (1425-1797).* Bruxelles et Paris: Vromant et Co., 1921.

Ward, Adolphus William. "The Effects of the Thirty Years' War," *Collected Papers: Historical, Literary, Travel and Miscellaneous,* Vol. I. Cambridge: At the University Press, 1921. Portrays the effects of the war on German university life.

Woodward, William Harrison. *Studies in Education During the Age of the Renaissance, 1400-1600.* Cambridge: University Press, 1924. Informative on the influence of humanism.

Znaniecki, Florian. *The Social Role of the Man of Knowledge.* New York: Columbia University Press, 1940. Though perhaps out of place in this bibliography, this work has provoked some fruitful reflection on the function and self-image of academics.

(I would like to stress what the reader may have already noted, that the above list contains no studies devoted specifically to the Italian universities after Galileo. I have been able to locate no such works, a rather remarkable lacuna in the literature of the universities. Considering their brilliance in the sixteenth century, particularly with regard to the natural sciences, the fate of the Italian universities in the seventeenth and eighteenth centuries would seem a subject of great promise, either as a study of continuing vitality or of the sources and methods of intellectual repression.)

Science and Philosophy in the Early Modern Period:

Boas, Marie. "The Establishment of the Mechanical Philosophy," *Osiris,* Vol. X (1952), pp. 412-541. Concentrates on atomism in the sixteenth and seventeenth centuries.

—. *The Scientific Renaissance, 1450-1630.* New York: Harper and Brothers, c. 1962.

Burtt, Edwin Arthur. *The Metaphysical Foundations of Modern Physical Science.* [Rev. ed.]; Garden City, N. Y.: Doubleday, 1932. See E. W. Strong for balance.

Butterfield, Herbert. *The Origins of Modern Science: 1300-1800.* [Rev. ed.]; New York: Macmillan, 1957.

Cassirer, Ernst. *The Philosophy of the Enlightenment.* Trans. Fritz C. A. Koelln and James P. Pettegrove; Princeton: Princeton University Press, 1951.

Castiglioni, Arturo. *A History of Medicine.* Trans. and ed. E. B. Krumbhaar. New York: Alfred A. Knopf, 1941.

Cohen, I. Bernard. *Franklin and Newton: An Inquiry into Speculative Newtonian Experimental Science and Franklin's Work in Electricity as an Example Thereof.* Philadelphia: The American Philosophical Society, 1956. Many helpful observations about eighteenth-century science.

Crombie, A. C. *Medieval and Early Modern Science.* [Rev. 2nd ed.]; 2 vols.; Garden City, N. Y.: Doubleday and Company, Inc., 1959.

Dijksterhuis, E. J. *The Mechanization of the World Picture.* Trans. C. Dikshoorn; Oxford: At the Clarendon Press, 1961.

Dillenberger, John. *Protestant Thought and Natural Science.* London: Collins, 1961.

Dugas, René. *Mechanics in the Seventeenth Century.* Trans. Freda Jacquot; Neuchatel, Switzerland: Éditions du Griffon; New York: Central Book Company, Inc.; c. 1958.

Ferguson, Allan, ed. *Natural Philosophy Through the 18th Century and Allied Topics.* Commemorative number of *The Philosophical Magazine*; London: Taylor and Francis, Ltd., 1948.

Gay, Peter. *The Enlightenment: An Interpretation.* 2 vols.; New York: Alfred A. Knopf, 1966-9.

Gilbert, Neal W. *Renaissance Concepts of Method.* New York: Columbia University Press, 1960.

Gillispie, Charles. *The Edge of Objectivity: An Essay in the History of Scientific Ideas.* Princeton, New Jersey: Princeton University Press, 1960.

Haldane, Elizabeth S. *Descartes: His Life and Times.* London: John Murray, 1905.
Hall, A. Rupert. "Merton Revisited, or Science and Society in the Seventeenth Century," *History of Science,* Vol. II (1963), pp. 1-16. A thoughtful reconsideration of arguments over the role of social and economic factors in the scientific revolution of the seventeenth century.
—. *The Scientific Revolution, 1500-1800: The Formation of the Modern Scientific Attitude.* London and New York: Longmans and Green, [1954]. A comprehensive survey.
Hesse, Mary B. *Forces and Fields: The Concept of Action at a Distance in the History of Physics.* London and New York: T. Nelson, [1961].
King, Lester S. *The Medical World of the Eighteenth Century.* Chicago: The University of Chicago Press, 1958.
Kline, Morris. *Mathematics and the Physical World.* New York: Crowell, [1959].
Koyré, Alexandre. *From the Closed World to the Infinite Universe.* Baltimore: Johns Hopkins Press, [1957]. By the brilliant scholar who has done most to make the history of science a history of philosophy.
—. "Galileo and Plato," *Roots of Scientific Thought.* Philip P. Wiener and Aaron Noland, eds.; New York: Basic Books, c. 1957.
—. "Galileo and the Scientific Revolution of the Seventeenth Century," *The Philosophical Review,* Vol. LII (1943), pp. 333-48.
—. "Influences of Philosophic Trends on the Formulation of Scientific Theories," *Scientific Monthly,* Vol. LXXX (1955), pp. 107-11.
—. *Newtonian Studies.* London: Chapman and Hall, 1965.
Kristeller, Paul Oskar. *Renaissance Thought: The Classic, Scholastic, and Humanist Strains.* New York, etc.: Harper and Row, 1961.
—. *Renaissance Thought II: Papers on Humanism and the Arts.* New York, etc.: Harper and Row, 1965.
Kuhn, Thomas S. *The Copernican Revolution.* Cambridge, Mass.: Harvard University Press, 1957.
—. *The Structure of Scientific Revolutions.* Chicago and London: The University of Chicago Press, 1962. Suggestive theorizing about the nature of scientific activity and change.
Laudan, Laurens. "Theories of Scientific Method from Plato to Mach: A Bibliographical Review," *History of Science,* Vol. VII (1968), pp. 1-63.
McKeon, Richard P. "Aristotle and the Origins of Science in the West," *Science and Civilization.* Robert C. Stauffer, ed.; Madison: University of Wisconsin Press, 1949.
—. "Aristotle's Conception of the Development and the Nature of Scientific Method," *Journal of the History of Ideas,* Vol. VIII (1947), pp. 3-44.
—. "Philosophy and the Development of Scientific Methods," *Journal of the History of Ideas,* Vol. XXVII (1966), pp. 3-22.
Mouy, Paul. *Le Développement de la physique cartésienne, 1646-1712.* Paris: J. Vrin, 1934.

Nordenskiöld, Erik. *The History of Biology*. Trans. Leonard Bucknall Eyre; New York and London: Alfred A. Knopf, 1928.

Popkin, Richard H. *The History of Scepticism from Erasmus to Descartes*. Rev. ed.; Assen: Van Gorcum en Comp. N. V., 1964.

Randall, John Herman, Jr. *Aristotle*. New York and London: Columbia University Press, c. 1960.

—. *The Career of Philosophy: From the Middle Ages to the Enlightenment*. Columbia University Press, c. 1962.

Schofield, Robert E. *Mechanism and Materialism, British Natural Philosophy in An Age of Reason*. Princeton, N. J.: Princeton University Press, 1970.

Serrurier, Cornelia. *Descartes, l'homme et le penseur*. Paris: Presses Universitaires de France; Amsterdam: Editions Françaises d'Amsterdam; [1951].

Smith, Norman Kemp. *New Studies in the Philosophy of Descartes*. New York: Russell and Russell, Inc., 1963.

Solmsen, Friedrich. *Aristotle's System of the Physical World: A Comparison with His Predecessors*. Ithaca, N. Y.: Cornell University Press, 1960. Clearly reveals the basic assumptions and concerns that shaped Aristotelian physics.

Spink, J. S. *French Free-Thought from Gassendi to Voltaire*. New York: Greenwood Press, c. 1960. Much on atomism as well as Cartesianism in French thought during the period.

Strong, E. W. *Procedures and Metaphysics, A Study in the Philosophy of Mathematical-Physical Science in the Sixteenth and Seventeenth Centuries*. Berkeley, Calif.: University of California Press, 1936. Questions the importance of metaphysical preconceptions in determining the beginnings of early modern science.

Westfall, Richard S. *Force in Newton's Physics: The Science of Dynamics in the Seventeenth Century*. London: Macdonald; New York: American Elsevier; 1971.

Whitehead, Alfred North. *Science and the Modern World*. New York: The Macmillan Company, 1925.

Wild, John. "The Cartesian Deformation of the Structure of Change and Its Influence on Modern Thought," *The Philosophical Review*, Vol. L (1941), pp. 36-59.

de Wulf, Maurice. *Scholasticism Old and New: An Introduction to Scholastic Philosophy, Medieval and Modern*. Trans. P. Coffey; Dublin: M. H. Gill and Son, Ltd.; London: Longmans, Green and Co., 1910.

PRIMARY SOURCES

Major Works:

Allamand, Jean Nicolas Sébastien. "Histoire de la vie et des ouvrages de Mr. 's Gravesande," *Oeuvres philosophiques et mathématiques de Mr. G. J. 's Gravesande*. Jean N. S. Allamand, ed.; 2 vols.; Amsterdam: Chez Marc Michel Rey, 1774.

Burgersdijck, Franco. *Collegium physicum*. 2nd ed.; Lugduni Batavorum: Ex officinâ Elziviriorum, 1642.

—. *Idea philosophiae naturalis, sive methodus definitionum et controversiarum physicarum*. 2nd ed.; Lugd. Batavorum: Ex officina Bonavent. et Abrahami Elzevir, 1627.

—. *Institutionum metaphysicarum libri duo*. Rev. ed.; Lugduni Batavorum: Apud Hieronymum de Vogel, 1642.

le Clerc, Jean. "Eloge de feu Mr. de Volder Professeur en Philosophie et aux Mathematiques, dans l'Academie de Leide," *Bibliotheque choisie, pour servir de suite a la Bibliotheque universelle* (Amsterdam: Chez Henri Schelte, 1703-13), t. XVIII, pp. 346-401.

Geulincx, Arnold. *Sämtliche Schriften in fünf Bänden*. H. J. de Vleeschauwer, ed.; 5 vols.; Stuttgart-Bad Cannstatt: Friedrich Frommann, 1965-68.

's Gravesande, Willem Jacob. *Oeuvres philosophiques et mathématiques*. Jean N. S. Allamand, ed.; 2 vols.; Chez Marc Michel Rey, 1774.

—. *Philosophiae Newtonianae institutiones, in usus academicos*. Lugduni Batavorum: Apud Petrum Vander Aa, 1723.

—. *Physices elementa mathematica, experimentis confirmata, sive introductio ad philosophiam Newtonianam*. 2 vols.; Lugduni Batavorum: Apud Petrum Vander Aa et B. et P. Janssonios Vander Aa, 1720-1.

von Haller, Albrecht. *Haller in Holland: Het Dagboek van Albrecht von Haller van zijn Verblijf in Holland (1725-1727)*. G. A. Lindeboom, ed.; Delft: Koninklijke Nederlandsche Gist- en Spiritusfabriek N. V., 1958.

Heereboord, Adriaan. *Meletemata philosophica, maximam partem, metaphysica*. Lugduni Batavorum: Ex officinâ Francisci Moyardi, 1654.

—. *Meletemata philosophica*. Rev. and expanded ed.; 2 vols.; Neomagi, Ex officina Andreae ab Hoogenhuysen, 1664-65.

—. *Philosophia naturalis, cum commentariis peripateticis antehac edita: nunc vero hac posthumâ editione mediam partem aucta, et novis commentariis, partim è Nob. D. Cartesio, Cl. Berigardo, H. Regio, aliisque praestantioribus philosophis, petitis, partim ex propria opinione dictatis, explicata*. Lugduni Batavorum: Ex officinâ Cornelii Driehuysen, 1663.

—. *Philosophia, naturalis, moralis, rationalis*. Lugduni Batavorum: Ex officinâ Francisci Moyardi, 1654.

Jacchaeus, Gilbertus. *Institutiones physicae*. Rev. ed.; Lugduni Batavorum: Excudebat Vidua Ioannis Patij, 1624.

Kyper, Albert. *Institutiones physicae*. 2 vols.; Apud Franciscum Moiardum et Adrianum Wyngaerden, 1645.

Molhuysen, P. C., ed. *Bronnen tot de Geschiedenis der Leidsche Universiteit*. 7 vols.; 's-Gravenhage: Martinus Nijhoff, 1913-24.

van Musschenbroek, Petrus. *Compendium physicae experimentalis, conscriptum in usus academicos*. Venetiis: Apud Franciscum ex Nicol. Pezzana, 1769.

—. *Dissertatio physica experimentalis de magnete*. Vienna: Typis Joannis Thomae Trattner, 1754.

—. *Elementa physicae, conscripta in usus academicos.* Lugduni Batavorum: Apud Samuelem Luchtmans, 1741.

—. *Essai de physique.* Trans. Pierre Massuet; Leyden: Chez Samuel Luchtmans, 1739.

—. *Institutiones physicae, conscriptae in usus academicos.* Lugduni Batavorum: Apud Samuelem Luchtmans et Filium, 1748.

—. *Introductio ad philosophiam naturalem.* Lugduni Batavorum: Apud Sam. et Joh. Luchtmans, 1762.

—. *Physicae experimentales, et geometricae, de magnete, tuborum capillarium vitreorumque speculorum attractione, magnitudine terrae, cohaerentia corporum firmorum dissertationes: ut et ephemerides meteorologicae ultrajectinae.* Lugduni Batavorum: Apud Samuelem Luchtmans, 1729.

—. "De viribus magneticis," *Philosophical Transactions,* Vol. XXXIII (for the years 1724-5; published 1726), pp. 370-7.

de Raey, Joannes. *Clavis philosophiae naturalis, seu introductio ad naturae contemplationem, Aristotelico-Cartesiana.* Lugduni Batavorum: Ex officinâ Johannis et Danielis Elsevier, 1654.

Senguerdius, Wolferdus. "Dictata in Renati Des-Cartes Principia Anno 1690." Unpublished manuscript in the Koninklijke Bibliotheek in The Hague.

—. *Inquisitiones experimentales.* 2nd ed.; Lugduni Batavorum: Apud Cornelium Boutesteyn, 1699.

—. *Philosophia naturalis, quatuor partibus primarias corporum species, affectiones, differentias, productiones, mutationes, et interitus, exhibens.* 2nd edition; Lugduni Batavorum: Apud Danielem à Gaesbeeck, 1685.

—. *Rationis atque experientiae connubium.* Roterodami: Apud Bernardum Bos, 1715.

von Uffenbach, Zacharias Konrad. *Merkwürdige Reisen durch Niedersachsen, Holland und Engelland.* 3 vols.; Ulm, 1754.

de Volder, Burchardus. *Disputationes philosophicae omnes contra atheos.* Medioburgi: Apud Johannem Lateranum, 1685.

—. *Disputationes philosophicae sive cogitationes rationales de rerum naturalium principiis.* Medioburgi: Typis Remigii Schreverii, 1681.

—. *Exercitationes academicae, quibus Ren. Cartesii philosophia defenditur adversus Petri Danielis Huetii Episcopi Suessionensis Censuram philosophiae Cartesianae.* Amstelaedami: Apud Arnoldum van Ravestein, 1695.

—. *Quaestiones academicae de aëris gravitate.* Medioburgi: Typis Viduae Remigii Schreverii, 1681.

[Watson, Elkanah]. *A Tour of Holland, in MDCCLXXXIV.* Worcester, Mass.: Isaiah Thomas, 1790.

Orations and Individual Student Disputations:
(The disputations are available in the University of Leiden Library unless otherwise indicated.)

Baumgartus, Valentinus. *Disputatio physica de motu prima,* sub praesidio D.

Burcheri de Volder. Lugduni Batavorum: Apud Viduam et Haeredes Johannis Elsevirii, 1671. [Universitätsbibliothek Erlangen-Nürnberg]

Bazin, Joannes Augustus. *Disputatio philosophica de formis,* praeside D. Jacobo Bernardo. Lugduni Batavorum: Apud Jacobum Poereep, 1713. [British Museum]

Boerhaave, Hermannus. "Oratio... de comparando certo in physicis," *Opuscula omnia, quae hactenus in lucem prodierunt.* Hagae-Comitis: Apud J. Neaulme, 1738.

Branchu, Balthazar. *Dissertatio philosophica de elementis,* sub praesidio Jacobi Bernardi. Lugduni Batavorum: Apud Petrum Vander Aa, 1715. [British Museum]

van Bronchorst, Henricus. *Disputatio philosophica de vera gravitatis causa,* sub praesidio D^{ni}. Burcheri de Volder. Lugduni Batavorum: Apud Abrahamum Elzevier, 1685. [National Széchényi Library, Budapest]

Bruno, Johannes. *Disputatio physica de motu, tertia et ultima,* sub praesidio D. Burcheri de Volder. Lugduni Batavorum: Apud Viduam et Haeredes Joannis Elsevirii, 1675. [Universitätsbibliothek Erlangen-Nürnberg]

Casembroot, Gysbertus Henricus. *Disputatio philosophica quae est de mundi systemate,* sub praesidio D. Burcheri de Volder. Lugduni Batavorum: Apud Abrahamum Elzevier, 1694.

Chardevenus, Gedeon. *Disputatio physica, de mundo,* sub praesidio D. Adriani Heereboord. Lugduni Batavorum: Ex officinâ Bonaventurae et Abrahami Elsevir, 1650. [Bibliotheek der Rijksuniversiteit te Utrecht]

vander Codde, Pontianus. *Disputatio philosophica de motu,* sub praesidio D. Burcheri de Volder. Lugduni Batavorum: Apud Abrahamum Elzevier, 1684. [National Széchényi Library, Budapest]

Derecskei, Paulus. *Exercitatio philosophica, coelorum, siderumque lucidorum originem, et phaenomena, methodo synthetico-mathematicâ demonstrans,* sub praesidio D. Burcheri de Volder. In four parts; Lugduni Batavorum: Apud Abrahamum Elzevier, 1682. [MTA Könyvtára, Bibliotheca Academiae Scientiarum Hungaricae, Budapest]

Erckelens, Jacobus. *Exercitationum philosophicarum tertia et vicesima, quae est de corpore,* sub praesidio D. Burcheri de Volder. Lugduni Batavorum: Abrahamum Elzevier, 1692.

Hendrix, Michaël. *Disputatio philosophica, continens positiones miscellaneas, nobiliores, ex universa philosophia depromptas,* sub praesidio D. Adriani Heereboord. Lugduni Batavorum: Ex officina Viduae Johannis du Pre, 1661.

le Keux, Philippus. *Disputatio philosophica, de fine,* sub praesidio D. Francisci du Ban. Lugduni Batavorum: Ex officina Bonaventurae et Abrahami Elsevir, 1640. [Bibliotheek der Rijksuniversiteit te Utrecht]

Koleseri, Samuel. *Disputatio philosophica inauguralis de systemate mundi,* pro gradu doctoratus, et liberalium artium magisterio. Lugduni Batavorum: Apud Viduam et Haeredes Johannis Elsevirii, 1681.

Kopeczi, Johannes. *Disputatio philosophica de cometis, secunda bipertita,* sub praesidio D. Johannis de Raei. Lugduni Batavorum: Apud Viduam et Haeredes Johannis Elsevirii, 1666.

von der Lahr, Paulus. *Disputatio philosophica de absoluta quiete,* sub praesidio D. Burcheri de Volder. Lugduni Batavorum: Abrahamum Elzevier, 1684. [National Széchényi Library, Budapest]

van Musschenbroek, Petrus. *Oratio de certa methodo philosophiae experimentalis.* Trajecti ad Rhenum: Apud Guilielmum Vande Water, 1723.

—. "Oratio de methodo instituendi experimenta physica," *Tentamina experimentorum naturalium captorum in Academia del Cimento.* Lugduni Batavorum: Apud Joan. et Herm. Verbeek, 1731.

Pauw, Hadrianus. *Disputatio philosophica, continens conclusiones aliquot, ex universa philosophia depromptas,* sub praesidio D. Francisci du Ban. Lugduni Batavorum: Ex officina Joannis Maire, 1637. [British Museum]

de Reus, Abrahamus. *Disputatio philosophica, de constitutione physicae,* sub praesidio D. Johannis de Raei. Lugduni Batavorum: Apud Viduam et Haeredes Johannis Elsevirii, 1668. [British Museum]

Rhijn-Diick, Jacobus. *Disputatio philosophica, continens positiones miscellaneas,* sub praesidio D. Adriani Heereboord. Lugduni Batavorum: Apud Johannem Elsevirium, 1659.

Schuyl, Hermannus. *Disputatio philosophica inauguralis de vi corporum elastica,* pro gradu doctoratus, nec-non liberalium artium magisterio. Lugduni Batavorum: Apud Abrahamum Elzevier, 1688.

vanden Velden, Henricus. *Disputatio de substantia, pars secunda,* sub praesidio D. Johannis de Raei. Lugduni Batavorum: Ex officina Abrahami à Geerevliet, 1659.

de Volder, Burchardus. *Disputatio medica, inauguralis, de natura,* pro gradu doctoratus, summisque in medicina honoribus. Lugduni Batavorum: Apud Severinum Matthiae, 1664.

—. *Oratio de conjungendis philosophicis et mathematicis disciplinis.* Lugduni Batavorum: Apud Jacobum Voorm, 1682.

—. *Oratio de novis et antiquis.* Lugduni Batavorum: Apud Cornelium Boutestein, 1709.

—. *Oratio de rationis viribus, et usu in scientiis.* 2nd ed.; Lugduni in Batavis: Apud Fredericum Haringium, 1698.

—. *Oratio qua ... sese laboribus academicis abdicavit.* Lugduni Batavorum: Apud Cornelium Boutestein, 1705.

de Witte van Schooten, Joannes Fransiscus. *Dissertatio philosophica inauguralis de solido, ejusque partium, nec non hemisphaeriorum concavorum, et cylindrorum solidorum cohaerentia,* pro gradu doctoratus, et magisterii, summisque in philosophia et artibus liberalibus honoribus, et privilegiis, rìte, ac legitimè consequendis. Lugduni Batavorum: Apud Viduam Cornelii Boutesteyn, 1712.

INDEX

Accident, accidents, 21–4, 51–2, 83, 86.
Action at a distance, 124, 127.
Air (vacuum) pump, 98, 100–1, 136–7.
Albinus, Bernhard Siegfried, 152.
Algebra, 135, 149.
Allamand, Jean Nicolas Sébastien, 117, 121–2, 133, 152–3.
Altdorf, University of, 6 (*n.* 23), 97.
Alteration, 27, 53–4.
Amsterdam, Athenaeum in, 73, 78.
Amsterdam, city of, 10 (*n.* 43).
Aquinas, Thomas, 21.
Archimedes, 104.
Aristotelianism (peripatetic philosophy), 12, 15, 17, 32–3, 36, 38–9, 44, 47, 63–4, 73, 75–6.
Aristotle, 12–3, 16, 18–20, 23, 25–8, 30, 33, 36, 38, 44, 51, 53, 55–6, 59–60, 63–7, 70, 97, 106.
Arminian movement, 15.
Astronomy (*see also* New astronomy), 29–30, 119.
Atheism, 36, 45, 143.
Atomism, atomists, 12, 65–6, 79–80, 95.
Attraction, attractions, 122, 124–7, 129.
Averroës, 53.
Avicenna, 53.
Bacon, Francis, 59–60, 89, 98, 100, 102.
du Ban, François, 37–9.
Barometric experiments, 70–1, 98–100.
Bérigard, Claude, 48, 53; *Circulis Pisanus*, 48.
Berlin, Academy at, 117.
Bernard, Jacques, 113 (*n.* 4), 114, 116.
Bible (Scripture), the, 41, 46, 77, 141.

Bodecher Benning, Johannes, 37.
Boerhaave, Hermannus, 7–8, 115–6, 119, 135, 152; *Elementa chemiae*, 115.
de le Boe Sylvius, Franciscus, *see* Sylvius, Franciscus de le Boe.
Bologna, University of, 97.
Bornius, Henricus, 46–7.
Boyle, Robert, 100–2, 104.
Brahe, Tycho (*see also* Tychonic system), 30, 42, 79, 141.
Brandenburg, Elector of, 9, 151.
Breda, Illustre School at, 40.
Burgersdijck, Franco, Chapt. II, 35, 37–43, 48–50, 53, 56, 64, 75, 84, 89, 109 (*n.* 89), 130–2, 137, 141, 143–5, 147; *Collegium physicum*, 16 ff., 32, 39; *Idea philosophiae naturalis*, 16.
Calculus, 110–1.
Calvinism (Reformed Religion), Calvinists, 2, 4, 35, 38, 43, 74–8.
Cambridge, University of, 2–3, 8–9, 79, 97, 150 (*n.* 33).
Campanella, Thomas, 59.
Cartesianism (*see also* Physics *and* Rationalism, Cartesian), 34–7, 39, 43–4, 47, 58, 61, 64, 71, 73–9, 90, 103, 111, 118, 140, 143.
Cause, causality (*see also* Gravity, gravitation *and* Motion), 17, 93–5, 127–31, 144–5.
Celestial substance, 27–8, 30–2, 40–1, 56–7, 66, 141.
Circulation of the blood, theory of, 7, 36, 39.
Clear and distinct ideas, 50, 83, 91–5, 107.

INDEX

le Clerc, Jean, 97, 101, 110-2, 148.
Clusius, Carolus, 6.
Coccejanism, 77.
Coccejus, Johannes, 75, 77-8.
Colepresse, Samuel, 73-4.
Comenius, Johannes Amos, 59.
Copernicus, Nicolaus, 30-1, 41-2, 147.
Copernicanism; see Heliocentrism.
Corpuscular imagery, 51-3, 58, 69-71, 79-80, 104, 121-2, 142.
Corruption, 18-23, 27, 30, 38, 40, 52-4, 57.
Cosmography, 116.
Cosmos, 28-30, 32, 40, 42-3, 56, 58, 71-2, 141-2.
Craanen, Theodorus, 74-6.
Crystalline orbs or spheres, 30, 40, 72.
Cunaeus, 133.
Curriculum, philosophy, see Philosophy curriculum.
Dematius, Carolus, 96 (*n.* 39), 97 (*n.* 40).
Derecskei, Paulus, 107.
Desaguliers, John Theophilus, 116, 118.
Descartes, René, 33-7, 44-6, 48, 50-1, 53-4, 56-64, 66-7, 69, 71, 77-82, 85, 89, 95, 98, 106, 112, 121-2, 141, 144, 147-8; *Discours de la méthode*, 33; *Meditationes de prima philosophia*, 33; *Principia philosophiae*, 33-4, 69, 71, 148.
Diderot, Denis, 152.
Disputation, disputations, 15-6, 36-7, 40, 44-7, 59, 64, 76, 86, 94, 96, 110.
Donne, John, 93.
Dort, Synod of, 15.
Douai, University of, 2 (*n.* 3).
Doubt, Cartesian, 34-5, 44, 75.
Duisburg, University of, 119.
Dutch Republic, see United Provinces.
Eclectic, 48, 147-8.
Edinburgh, University of, 149, 151-2.
Electricity, 122, 132-3, 137, 139, 153 (*n.* 48).
Elements: peripatetic, 17, 27-30, 32, 40-1, 53, 57; Cartesian, 57, 70.
Empiricism, see Sense perception.
Encyclopédie, 152.
Engineering, 10 (*n.* 46), 117.
Enlightenment, the, 144.
Ethics, 5, 14-5, 44, 59, 73, 114, 117, 143.

Experience, see Sense perception.
Experimentation, experiments, 43, 58, 60, 71, 79, 96-106, 112-4, 118, 121, 131-8, 148-9.
Extension (Cartesian body), 50-1, 58, 64-6, 70-1, 79, 81-5, 103, 107, 142.
Fabricius ab Aquapendente, Hieronymus, 6.
Fire, 28, 41, 57, 122, 125.
Fluidity, 54, 70, 125, 137.
Force, forces, 23, 25-6, 122-7, 130, 132-4, 136.
Form, 17, 19-22, 26, 29, 31, 35, 50-2, 55, 57, 71, 79.
Franeker, University of, 3, 97, 111 (*n.* 98).
Frederick II, 117.
Fullenius, Bernhardus, 111 (*n.* 98).
Galileo Galilei, 30-2, 34, 42, 100-1, 104, 107-8; *Dialogo sopra i due massimi sistemi del mondo, Tolemaico, e Copernicano*, 32-3; *Discorsi e dimostrazioni matematiche, intorno à due nuove scienze*, 33.
Gaubius, Hieronymus David, 152.
Generation, 17-23, 27, 30, 38, 40, 51-4, 57.
Geneva, Academy at, 2 (*n.* 3), 78.
Geometry, 106-8, 121, 131-2, 135, 149.
George I, 116.
Geulincx, Arnold, 73-5, 82, 84-5, 87.
God, 41, 45, 47, 66-7, 70, 75, 80-2, 84, 87, 128-30, 143-4.
Goedaert, Johannes, 102.
der Goes, 77.
Golius, Jacobus, 44, 63.
Göttingen, University of, 151-2.
's Gravesande, Willem Jacob, 114, 116 ff., 142-4, 148-50, 152-3; *Essai de perspective*, 116; "Essai d'une nouvelle théorie sur le choc des corps," 134; "Mathematical Demonstration of the Care God Takes to Direct that which Takes Place in this World, Drawn from the Number of Boys and Girls Born Daily," 143; *Philosophiae Newtonianae institutiones*, 117-8; *Physices elementa mathematica*, 117-8, 120, 123, 131, 135-6.
Gravity, gravitation, 40, 42, 80-1, 109, 123-6, 129; cause of, 80-1, 109, 124.

INDEX

Groningen, University of, 3, 97.
von Guericke, Otto, 98, 101–2.
Haarlem, Society at, 150 (*n*. 33).
Hague, The, 75, 116.
Halle, University of, 9, 151.
von Haller, Albrecht, 135, 152.
Harderwijk, University of, 3.
Hart Hall, 97.
Harvey, William, 7, 36, 39, 102.
Heereboord Adriaan, 37–9, 44–60, 62–6, 69, 98, 138, 141–4, 146–7; *Philosophia naturalis cum novis commentariis explicata*, 48 ff.; *Philosophia, naturalis moralis, rationalis*, 48.
Heidanus, Abraham, 73–5, 77–8, 140; *Consideratiën over eenige Saecken onlanghs voorgevallen in de Universiteyt binnen Leyden*, 78.
Heliocentrism, 30–2, 35, 41–2, 56–7, 69, 77, 79–80, 141–2, 147.
d'Holbach, Paul Henri Thiry, Baron, 152.
Holland, province of, 2.
Hulsius, Antonius, 77.
Huygens, Christiaan, 101, 109, 111–2.
Hypotheses, 105, 121–3, 125, 131.
Illustre scholen, 3, 40, 97.
Impetus, 25, 56, 66, 80–1.
Incomprehensibility and inexplicability in physics, 122–5, 127–31, 133–4, 138, 142, 144.
Infinity and "indefinite" expanse, 58, 77, 81–4, 87, 123.
Instruments, scientific and experimental, 96, 100, 113, 119, 133–4, 136–7, 149, 153 (*n*. 48).
Intelligentia, 90–1.
Jacchaeus, Gilbertus, 14–5, 84; *Institutiones physicae*, 14–5.
Jena, University of, 7.
Journal des sçavans, 118.
Journal littéraire, 116, 134.
Keill, John, 118.
Kyper, Albert, 39–43, 49, 58, 64, 79–80, 83, 98, 141, 146–7; *Institutiones physicae*, 39–43.
La Flèche, 37.
de La Mettrie, Julien Offray, 152.
van Lansbergen, Philippus, 31.
Law, 5, 7, 11, 116, 140.
Laws of nature, 67–8, 104, 128–30, 132; of gravitation, 130; of hydrostatics, 104; of impact, 68, 87, 106, 110, 124, 137; of motion, 123, 130, 132.
van Leeuwenhoek, Antony, 102.
Leibniz, Gottfried Wilhelm, 111 (*n*. 99), 118, 120, 134.
Leiden, city of, 2, 9–10, 35, 75 (*n*. 11), 153.
Leiden, University of: astronomy instruction at, 6, 116, 130, 137, 152 (*n*. 48); Cartesianism at, 36–7, 39, 43–4, 47, 61–2, 73–8, 131, 144; chemical instruction at, 6–8, 96 (*n*. 39), 115; demonstration hall for experimental physics, 96, 113, 116–7, 136; experimentalism at, 14, 43, 96–8, 101–3, 113–7, 119, 131, 136, 138, 144, 150–1, 153 (*n*. 48); faculty of medicine, 7–8, 39, 63, 76, 115, 152; faculty of law, 7; faculty of philosophy, or arts, 14–5, 46, 74, 76, 153–4; faculty of theology, 43, 46–7, 74–8; foreign students at, 4; founding of, 2–3; innovations and new facilities at, 6–7, 96–7, 101, 103, 108, 114, 141; internal organization, 4–5; mathematical instruction at, 10 (*n*. 46), 14, 110, 135; mathematical physics at, 108, 110, 131–2, 135, 138, 144; medical instruction at, 6–8, 115, 151 (*n*. 34), 152; the new science at, 10, 13, 139–41, 144, 146, 150–1, 153–4; observatory at, 6, 116, 152 (*n*. 48); philosophical restrictions at, 44–6, 61–2, 76–7, 140–1, 154; philosophy instruction at, 5, 14–6, 96–8; professorship of astronomy at, 114, 116, 152; professorship of mathematics at, 14, 44, 108, 114, 116, 119, 153 (*n*. 48); professorship of physics at, 14–6, 37, 39–40, 43, 60, 62–3; religious toleration at, 4–5, 43, 74; state college for theology students at, 44, 75–6; stature in Europe, 3, 8–9, 113, 152, 154; student numbers at, 3–4; student turbulence at, 4, 45–6, 61–2, 76, 87.
Leyden jar, 132–3.
Light, 31, 70, 90, 122, 125, 132.
Lipsius, Justus, 5–6.
Logic, 5, 14–5, 37–8, 44, 73, 75, 117, 143.
Louis XIV, 74, 153.
Louvain, University of, 2, 6, 73.
Lucretius, 12.

Lulofs, Johannes, 120, 152–3.
Machine, machines, 101, 123, 136–8, 149.
Magdeburg hemispheres, 98 (including n. 49), 102, 104.
Magirus, Joannes, 39; *Physiologia Peripatetica*, 39.
Magnetism, 70, 125, 133–4.
Mathematics, 5, 14, 24, 91, 93, 106–7, 114, 116, 119; *re* physics and the new science, 24–5, 50, 58, 68, 71, 105–10, 112, 128, 131–2, 134–5, 138, 144–5.
Matter, 64, 123, 128; Cartesian (*see also* Extension), 52, 64–5, 70, 83; peripatetic, 17, 19–22, 49–51, 57, 79.
Maurice, Prince of Orange, 10 (n. 46), 15, 117.
Mechanical philosophy, 21, 25, 49–52, 80, 94, 101, 112, 130, 138, 142.
Mechanics, 124, 137.
Medicine, 5–8, 11, 36, 40, 43, 45, 62–3, 74, 105, 119, 140, 151 (n. 34), 152.
Metaphysics, 5, 15, 19, 24, 29, 44, 47–8, 62, 73, 75, 77, 93, 112, 117, 143.
Meteorology, 17, 137 (n. 137).
Method: pedagogical, 16–7, 32, 38, 43, 89, 135–6, 147–9; philosophic, 89, 131; scientific, 89, 100, 131.
Microscope, 102, 137.
Mobile, 26, 55.
Motion, 64, 87–8, 125–7; causality of, 24–6, 54–6, 65–7, 80–1, 125–7; corpuscular, 51–4; local motion, 27–9, 53–5, 86; natural motion, 23–7, 29, 32, 54–5, 66; new understanding of, 23, 25, 42, 53–6, 58, 64–9, 81, 84–7, 123, 125–8, 130, 132, 142–3, 145; quantity of, 68, 87; relativity of, 26, 67–9, 84–7, 125, 143, 145; traditional understanding of, 17, 22–9, 32, 54–5, 66, 80–1, 85–7, 142–3; violent motion, 25, 54.
Movens, 26, 55.
Münster, Peace of, 3.
van Musschenbroek, Jan, 119, 136.
van Musschenbroek, Petrus, 116, 119 ff., 142–5, 149–53; *Beginselen der Natuurkunde*, 120; *Compendium physicae experimentalis*, 120; *Cours de physique experimentale et mathematique*, 120; *Elementa physicae*, 120, 131; *Epitome elementorum physico-mathematicorum*, 119; *Essai de physique*, 120; *Institutiones physicae*, 120; *Introductio ad philosophiam naturalem*, 120, 144–5; *Physicae experimentales, et geometricae . . . dissertationes*, 119–20.
Natural body, 18–24, 26–9, 31, 49, 52, 56, 66, 71–2.
Nature, 13, 23, 59–60, 68–72, 87, 90–3, 100–1, 112, 121, 123, 127–8, 130, 132, 137–8, 141, 144–5, 148; peripatetic understanding of, 17, 24, 26, 29, 55, 72.
Navarre, College of, 97.
New astronomy, the, 29–32, 57.
New philosophy, the, 34–6, 47, 53, 61–2, 68, 78, 87, 140, 144.
New science, the, 10, 12–3, 17, 21–4, 27, 29, 32, 34, 49–50, 53, 57–8, 60, 68, 71, 88, 90, 103, 105, 111–2, 141–2, 144–7, 149.
Newton, Isaac, 2, 23, 79–80, 85, 97, 100, 107, 109–12, 115–6, 119–24, 126, 143, 148; *Philosophiae naturalis principia mathematica*, 109–12.
Newtonian philosophy, Newtonianism, 117, 120–1, 143; opposition to, 118.
Newtonian science, 88, 115, 118, 120, 126–7, 148.
Newtonians: Dutch, 118, 120–1; British, 134.
Nouvelles de la république des lettres, 113 (n. 4), 114.
Observation, *see* Sense perception.
van Oldenbarnevelt, Johan, 15.
Oldenburg, Henry, 73–4.
Oligarchy, Dutch, 10, 15, 74–5, 154.
Optics, 14, 137.
Orange, House of, 15.
Orange, Prince of (*see also* Maurice, William I, *and* William III), 40.
Order in nature, 29, 42–3, 68, 129, 144, 148.
Ordo secundum quem deinceps in Academia Leidensi Philosophia docebitur, 38–9.
Oxford, University of, 2–3, 8–9, 12, 97, 150 (n. 33).
Padua, University of, 2, 6–7.
Paris, *Académie des sciences* at, 133, 150 (including n. 33).

INDEX

Paris, University of, 3, 8, 11 (*n.* 49), 12, 97, 150 (*n.* 33).
Pascal, Blaise, 93, 99–101.
Patrizzi, Francesco, 59.
Pauw, Petrus, 6.
Pedagogy, *re* science and research, 15–7, 102–3, 135, 146–9.
Peripatetic philosophy, *see* Aristotelianism.
Peter I, 117.
Phenomena, 17, 60, 69–72, 80, 92, 94–5, 103, 106–8, 121, 124–5, 128–32, 138, 145.
Philip II, 2–3.
Philosophy curriculum, 5, 11–2, 110, 140, 146, 150–1.
Philosophy degrees, 11, 62, 74 (*including n.* 8), 78, 119, 150.
Physics: as a discipline, 5, 10–2, 14–7, 36, 38, 64, 135–7, 139–40, 144, 146–50; as a science, 24, 72, 92–6, 100, 104–10, 112, 125, 127–8, 130–2, 136, 138–9, 143–6; Cartesian (*see also* Rationalism, Cartesian), 33–6, 48, 50–1, 58, 64, 72, 75, 78–9, 87, 94, 101, 103, 112, 144, 148; modern (*see also* Science), 134; Newtonian, *see* Newtonian science; peripatetic, 15, 17–8, 23–5, 29–30, 32, 38–9, 41, 48–9, 71, 106.
Piety, 143–4.
Place, 17, 27, 72, 85–6, 125; natural places, 28–9.
Plenum, Cartesian, 84, 104.
Pneumatica, 114.
Pneumatics, 84, 98, 104.
Potentiality, 19–20, 24, 29, 51–2.
Praecognita, 64, 68–9.
Primary qualities, peripatetic, 41, 54, 57.
Principles, philosophic, 64, 79, 89, 93–5, 107, 130–1, 145, 147; Cartesian, 64, 68, 72, 94–6, 107, 131, 142, 144; peripatetic, 17, 18–23, 29, 49, 58, 71, 79, 142.
Privation, 17, 19–20, 24, 27, 49–50.
Ptolemaic system, 31–3, 42.
Ptolemy, 31.
de Raey, Johannes, Chapt. IV, 10 (*n.* 45), 45, 56, 60, 73–4, 81, 84, 90–1, 94, 98–100, 106, 122, 143–4, 147, 150; *Clavis philosophiae naturalis*, 63 ff., 85, 90, 94, 98–100, 106.
Ramus, Petrus, 59.
Rationalism, Cartesian, 35, 89–96, 103–8, 110–2, 123, 132.
Redi, Francesco, 102.
Reformation, the, 2, 12.
Reformed Church, Dutch, 15, 76–8.
Reformed Religion, *see* Calvinism.
Refraction, 14, 57, 125.
Regius (Henri de Roy), 35–6, 48, 53–4, 56, 62, 146.
Renaissance, the, 12.
Repulsion, repulsions, 125–6, 129.
Revius, Jacobus, 44–5, 62.
de Roberval, Gilles Personne, 99, 101.
Rohault, Jacques, 103, 112; *Traité de physique*, 103.
de Roy, Henri, *see* Regius.
Royal Society of London, 97, 101, 116, 150 (*including n.* 33).
St. Petersburg, Academy at, 117–8.
Salamanca, University of, 97.
Saumur, Academy at, 15.
Scaliger, Josephus Justus, 6.
Scepticism, sceptics, 35, 89–90, 92.
van Schooten, family, 10 (*n.* 46).
Schott, Gaspar, 102.
Schuyl, Hermannus, 110.
Science (*see also* Physics): conception of, 109–10, 130–1, 142, 144–6; medieval, 12, 149–50; modern, 10, 140, 144–6, 149, 154; the new science, *see* New science; Newtonian, *see* Newtonian science.
Scotists, 50.
Scripture, the, *see* Bible.
Sedan, Academy at, 43.
Senguerdius, Wolferdus, 76, 78–87, 92, 96, 98, 100–4, 111, 113–6, 122, 136–7, 141–2, 144, 148; *Inquisitiones experimentales*, 92, 98; *Philosophia naturalis*, 79 ff., 92, 100; *Rationis atque experientiae connubium*, 92, 100–1, 104, 114.
Sennertus, Daniel, 39; *Philosophia naturalis*, 39.
Sense perception (empiricism, experience, observation), 17, 40, 60, 71, 89, 90, 92–3, 95–6, 98–9, 101–5, 123, 125, 128, 131, 143.

Snellius, Willebrord, 14–5.
Solar system, 69, 123–4, 130, 137.
Space, 29, 69, 72, 82–7, 92, 104, 122, 125–6, 145; empty space, *see* vacuum.
Spanheim, Frederick, 75, 77.
States of Holland and Westfriesland, 4–5, 46–7, 61, 76.
Stevin, Simon, 10 (*n.* 46), 104.
Stuart, Adam, 43–8, 56, 60, 62.
Sublunar world, 27–8, 30–2, 40–1, 64, 70.
Substance, 21–2, 50, 52, 65, 72, 79, 83, 85, 87–8, 103, 107, 123–4, 127, 130; celestial, *see* Celestial substance.
Subtle matter, celestial fluid, or celestial matter, 64, 69–70, 79–80, 99, 104, 122–4.
Swammerdam, Jan, 102, 151 (*n.* 34).
Sylvius, Franciscus de le Boe, 7.
System, philosophic, 17, 32, 71, 90, 130, 135–6, 147–8.
Teleology, 26, 55, 68, 143.
Telescope, 30, 57, 137.
Telesio, Bernardino, 59.
Terminus a quo, 26, 55.
Terminus ad quem, 26–7, 29, 55, 80, 86.
Theology, 4–5, 9, 11, 14, 34–6, 40, 43–7, 60, 75, 77, 140, 143–4.
Thysius, Antonius, 46–7.
Time, 17, 122.
Torricelli, Evangelista, 71, 98–9, 101.
Triglandius, Jacobus, 44–5.
Trinity College, 97.
Tychonic system, 42, 79, 141.
von Uffenbach, Zacharias, 113–4.
United Provinces (Dutch Republic), 3, 9–10, 34–5, 37, 74, 97, 118–9, 153.
Universe (*see also* World), 27, 29, 34, 41–2, 87, 93, 95, 141, 143–4, 148; Cartesian, 57–8, 69, 72, 106, 141; Newtonian, 86, 123–4, 126, 128–9.
Universities (*see individual universities as well*): Dutch, 11, 35; English, 11 (*n.* 48), 16; German, 4, 9, 151–2; Italian, 2, 6, 8, 11 (*n.* 48), 97, 161–2; Scottish, 4, 9; Spanish, 8; *re* Aristotelianism, 12, 15, 33; *re* Cartesianism, 33, 89, 108; *re* the new science, 1–2, 10–1, 13, 15, 97, 101, 108, 113–4, 146, 149–51.
Utrecht, University of, 3, 35–6, 38–9, 47, 62, 74, 78, 119, 121, 131.
Vacuum (empty space, void), 17, 65, 83–7, 92, 98–9, 104, 123, 126, 137.
Vacuum pump, *see* air pump.
Vis (*see also* Force, forces), 126–7.
Vis insita (*see also* Force, forces), 126.
Vivés, Luis, 59.
Voetius, Gisbertus, 36.
Void, *see* Vacuum.
de Volder, Burchardus, 4 (*n.* 14), 74, 77–8, 79 (*n.* 37), 80–3, 85–7, 91–8, 101–5, 131, 135–6, 140, 144, 147–8; *Disputationes philosophicae sive cogitationes rationales de rerum naturalium principiis*, 94, 96; "The Powers of Reason," 112; *Quaestiones academicae de aëris gravitate*, 96, 98, 102, 104, 110.
Voltaire, 118; *Élémens de la philosophie de Newton*, 118.
Vortices, Cartesian, 57–8, 69, 71–2, 79–80.
Vossius, Isaac, 110–1 (*n.* 96).
de Vries, Gerardus, 76.
Walcheren, 77.
de Wale, Johan, 7.
Watson, Elkanah, 153.
West India Company, 37.
Wilhelmius, Wilhelmus, 76.
William I, Prince of Orange, 2–3.
William III, Prince of Orange, 75, 77.
Willis, Thomas (?), 102.
Whitehead, Alfred North, 13.
de Witt, Johan, 47, 75.
Wittichius, Christophorus, 74–5, 77–8.
Wittichius, Jacobus, 114, 117, 119.
Wolff, Christian, 118, 120.
World (*see also* Universe), 17–8, 27, 32, 81–2, 84, 107, 147.
Würzburg, University of, 97.
Zabarella, Giacomo, 53.

RAYMOND H. FOGLER LIBRARY
DATE DUE